ADVICE TO A YOUNG SCIENTIST

For Cameron
w/ love from
— Mammu
Nov 2014

BOOKS IN THE ALFRED P. SLOAN FOUNDATION SERIES

Disturbing the Universe *by Freeman Dyson*

Advice to a Young Scientist *by Peter Medawar*

THIS BOOK IS PUBLISHED AS PART
OF AN ALFRED P. SLOAN FOUNDATION PROGRAM.

P. B. MEDAWAR

ADVICE TO A YOUNG SCIENTIST

A Member of the
Perseus Books Group

Portions of this work originally appeared in *Harper's Magazine* and *The Sciences.*

Library of Congress Cataloging in Publication Data
Medawar, Peter Brian.
Advice to a young scientist.
1. Science—Vocational guidance. I. Title.
Q147.M38 1979 502'.3 79-1676

ISBN 0-465-00092-4 (paper)

Printed in the United States of America
Designed by Sidney Feinberg

To
The Royal Society
of London

Contents

Preface to the Series

The Alfred P. Sloan Foundation has for many years included in its areas of interest the encouragement of a public understanding of science. It is an area in which it is most difficult to spend money effectively. Science in this century has become a complex endeavor. Scientific statements are embedded in a context that may look back over as many as four centuries of cunning experiment and elaborate theory; they are as likely as not to be expressible only in the language of advanced mathematics. The goal of a general public understanding of science, which may have been reasonable a hundred years ago, is perhaps by now chimerical.

Yet an understanding of the scientific enterprise, as distinct from the data and concepts and theories of science itself, is certainly within the grasp of us all. It is, after all, an enterprise conducted by men and women who might be our neighbors, going to and from their workplaces day by day, stimulated by hopes and purposes that are common to all of us, rewarded as most of us are by occasional successes and distressed by occasional setbacks. It is an enterprise with its own rules and customs, but an understanding of that enterprise is accessible to any of us, for it is quintessentially human. And an understanding of the enterprise inevitably brings with it some insight into the nature of its products.

Accordingly, the Sloan Foundation has set out to encourage a representative selection of accomplished and articulate scientists to set down their own accounts of their lives in science. The form those accounts will take has been left in each instance to the author: one may choose an autobiographical approach, another may produce a coherent series of essays, a third may tell the tale of a scientific community of which he was a member. Each author is a man or woman of outstanding accomplishment in his or her field. The word "science" is not construed narrowly: it includes such disciplines as economics and anthropology as much as it includes physics and chemistry and biology.

The Foundation's role has been to organize the program and to provide the financial support necessary to bring manuscripts to completion. The Foundation wishes to express its appreciation of the great and continuing contribution made to the program by its Advisory Committee chaired by Dr. Robert Sinsheimer, Chancellor of the University of California—Santa Cruz, and comprising Dr. Howard H. Hiatt, Dean of the Harvard School of Public Health; Dr. Mark Kac, Professor of Mathematics at Rockefeller University; Dr. Daniel McFadden, Professor of Economics at the Massachusetts Institute of Technology; Robert K. Merton, University Professor, Columbia University; Dr. George A. Miller, Professor of Experimental Psychology at Rockefeller University; Professor Philip Morrison of the Massachusetts Institute of Technology; Dr. Frederick E. Terman, Provost Emeritus, Stanford University; for the Foundation, Arthur L. Singer, Jr., and Stephen White; for Harper & Row, Winthrop Knowlton and Simon Michael Bessie.

—ALBERT REES
President, Alfred P. Sloan Foundation

Author's Preface

I have tried to write the kind of book I myself should have liked to have read when I began research before most of my readers were born—that is not a patronizing comment but a straightforward recognition of the fact that most scientists are young in years and that no one actively engaged in research ever thinks of himself as old.

I am properly conscious, too, of joining the company of Polonius, Lord Chesterfield, and William Cobbett,[1] all well known for having advised the young. Although none of their advice was addressed to young scientists, some of it applies. The advice of Polonius was mainly prudential in character and though one can sense Laertes's haste to be away ("Most humbly do I take my leave, my Lord"), it is excellent advice.

Chesterfield's advice had mainly to do with manners, especially the arts of ingratiation. It has little relevance to the circles in which scientists move, which is perhaps just as well because it received a stunning blow from the tail of the great Leviathan of English letters. Chesterfield, Dr. Johnson declared, taught the manners of a dancing master and the morals of a whore.

Cobbett's advice was in a wide sense moral, though it had to do with manners too. Although Cobbett had not Dr. John-

1. William Shakespeare (1603), *Hamlet,* act 1, scene 3; Philip Dormer Stanhope, IV, Earl of Chesterfield (1694–1773), *Letters to His Son* (1774); William Cobbett (1763–1835), *Advice to Young Men and (incidentally) to Young Women* (1829).

son's formidable strength of mind there is as much good sense in a paragraph of Cobbett as there is in any other paragraph of English prose. The eyes of one or another or all three will be found to glare from these pages at the appropriate places, for it is hardly possible to write a book of advice without being influenced by what they had to say.

The scope and purpose of this little book are explained in the Introduction: it is not for scientists only but for all who are engaged in exploratory activities. And it is not only for the young in years; with no thought of extra charge, author and publisher resolved to include a few paragraphs of advice to older scientists too. I have had in mind another audience, as well: nonscientists who may for any reason be curious about the delights and vexations of being a scientist, or about the motives, moods and mores of members of the profession.

Any passage in this book that a reader may think especially apt and illuminating is that which was written for him or her; that which is well understood already will not be thought interesting and will pass by unnoticed.

I have been embarrassed throughout by the lack in English of an epicene personal pronoun or possessive adjective, so for the most part "he" will have to do for "she," and "his" for "hers." Chapter 5 will make it clear that everything I say applies to women if it applies to men.

Almost inevitably, this book embodies a personal "philosophy" having to do with the place of science and scientists in the world. It is a very opinionated book, so something more is needed by way of apologia. In wartime Britain, to establish a personal relationship with the public, the radio newsreaders always announced their identity, often in the following words: "This is the nine o'clock news and this is Stuart Hibberd reading it." Of the style and contents of this book I shall say only, "These are my opinions, and this is me giving them." I use the word *opinion* to make it clear that my judgments are not validated by systematic sociological research and are not hypotheses that have already stood up to repeated critical assaults. They are merely personal judgments, though I hope that some of them will be picked up by sociologists of science for proper investigation.

The experience that justifies my writing a book such as this is the following. I was for a good many years a tutor at Oxford in the days when a single tutor was wholly responsible for the intellectual upbringing of his pupils—an exciting enterprise for both parties. A good tutor taught the whole of his subject and not just that part of it in which he himself happened to be especially interested or proficient; to "teach" did not, of course, mean to "impart factual information," a relatively unimportant consideration, but rather to guide thought and reading and encourage reflection. I later became the head of university teaching departments, first in the University of Birmingham and latterly in University College London. After this I became the head of the National Institute for Medical Research, a large medical research institute populated by scientists of all ages and degrees of seniority.

In these environments I observed with great interest what was going on around me. Furthermore, I myself was young once.

Laying now the trumpet aside, I should like to express my indebtedness to my patrons, the Alfred P. Sloan Foundation, who made it so easily possible and so agreeable to fit the writing of this book into a busy professional life. It was my patrons' wish, not mine, that to caution or exemplify, I should draw upon my personal experience as a scientist more often than I was inclined to do.

The special circumstances of my life are such that no writing upon any subject would have been possible without the support and companionship of my wife. Although this particular book is a solo effort, my wife too has read it because I have come to have complete confidence in her ear and literary judgment.

The work of preparing the text for publication was that of my secretary and assistant, Mrs. Heys.

I should like also very specially to thank some close friends for their hospitality and patient forbearance while I was writing or dictating this book: Jean and Friedrich Deinhardt, Barbara and Oliver Poole, and Pamela and Ian McAdam.

P. B. MEDAWAR

1

Introduction

In this book I interpret "science" pretty broadly to refer to all exploratory activities of which the purpose is to come to a better understanding of the natural world. This exploratory activity is called "research," and research is my chief topic, although it is only a small fraction of the multitude of scientific or science-based activities, which include scientific administration, scientific journalism (which grows in importance with science itself), the teaching of science, the supervision and often the execution of many industrial procedures, especially in respect of drugs, prepared foods, machinery and other manufactures, and textiles and materials generally.

In America, 493,000 people classified themselves in a recent census as scientists,[1] a very large number even when whittled down to 313,000 by applying the more exacting taxonomic criteria adopted by the National Science Foundation. The numbers in Great Britain are about the same in proportion to total population. The Department of Industry reported that the total stock of qualified scientists in Great Britain in 1976 was 307,000, of whom 228,000 were described as "economically active." Ten years before, the corresponding figures had been 175,000 and 42,000. The number of scientists in the world considered as a whole must be between 750,000 and 1,000,000. Most are still

1. I quote these figures from Harriet Zuckerman's *Scientific Elite* (London: Macmillan, 1977).

young, and all are, or at one time were, in need of advice.

I make no apology for concentrating mainly on research. I do so in the same spirit as that in which the author of "Advice to Young Writers" would preoccupy himself with imaginative writing rather than with ancillary and supportive activities such as printing, publishing, or reviewing—important though they are. Although research in the natural sciences is my principal theme, I shall always be thinking of exploratory activities in general, and believe that what I say will bear upon sociology, anthropology, archaeology, and the "behavioral sciences" generally, and not just upon the world of laboratories, test tubes, and microscopes, for I am not forgetting that human beings are among the most prominent fauna of that "natural world" of which I said that it was our purpose to seek an understanding.

It is not easy and will not always be necessary to draw a sharp distinction between "real" research scientists and those who carry out scientific operations apparently by rote. Among those half-million or so practitioners who classified themselves as scientists might easily have been the kind of man employed by any large and well-regulated public swimming pool: the man who checks the hydrogen-iron concentration of the water and keeps an eye on the bacterial and fungal flora. I can almost hear the contemptuous snort with which the pretensions of such a one to be thought a scientist will be dismissed.

But wait; scientist is as scientist does. If the attendant is intelligent and ambitious, he may build upon his school science by trying to bone up a little bacteriology or medical mycology in a public library or at night school, where he will certainly learn that the warmth and wetness that make the swimming pool agreeable to human beings are also conducive to the growth of microorganisms. Conversely, the chlorine that discourages bacteria is equally offensive to human beings; the attendant's thoughts might easily turn to the problem of how best to keep down the bacteria and the fungi without enormous cost to his employer and without frightening his patrons away. Perhaps he will experiment on a small scale in his evaluation of alternative methods of purification. He will in any case keep a record of the relationship between the density of the population of microorganisms and the number of users of the pool, and experiment with adjusting the concentration of chlorine in accordance with his expectation of the number of his patrons on any particular day. If he does these things, he will be acting as a scientist rather than as a hired hand. The important thing is the inclination to get at the truth of matters as far as he is able and to take the steps that will make it reasonably likely he will do so. For this reason I shall not always make a distinction—and certainly never a class distinction (see Chapter 6)—between "pure" and applied science, a subject almost irremediably confused by a misunderstanding of the word *pure.*

In science a beginner will certainly read or be told "The scientist *this*" or "The scientist *that.*" Let him not believe it. There is no such person as *the* scientist. There are scientists, to be sure, and they are a collection as various in temperament as physicians, lawyers, clergymen, attorneys, or swimming-pool attendants. In my book *The Art of the Soluble* I put it thus:

> Scientists are people of very dissimilar temperaments doing different things in very different ways. Among scientists are collectors, classifiers and compulsive tidiers-up; many are detectives by temperament and many are explorers; some are artists and others artisans. There are poet-scientists and philosopher-scientists and even a few mystics. What sort of mind or temperament can all these people be supposed to have in common? *Obligative* scientists must be very rare, and most people who are in fact scientists could easily have been something else instead.

I remember saying of the dramatis personae in the story of the unraveling of the crystalline structure of DNA[2] that it would be hard even to imagine a collection of people more different from each other in origin and education, manner, manners, appearance, style, and worldly purposes than James Watson, Francis Crick, Lawrence Bragg, Rosalind Franklin and Linus Pauling.

I used the word *mystic* to refer to those few scientists who derive a perverse satisfaction from knowing that something is not known and who use that ignorance as a pretext for bursting out of the cruel confinements of positivism into the domain of

---

2. See P. B. Medawar, "Lucky Jim," in *The Hope of Progress* (London: Wildwood House, 1974).

rhapsodic intellection; but I am ashamed to say that after "and even a few mystics" I should now add "and even a few crooks."

The most crooked scientist I know is one who plagiarized a number of photographs and several paragraphs of text from a fellow worker and included them in a prize essay put up for competition by a college in one of the older universities. One of his judges was the man from whom his material had been stolen. A terrific row followed, but, luckily for the culprit, the body that employed him was anxious above all else to avoid any public scandal. The culprit was accordingly "redeployed" into another scientific institution and has pursued a moderately successful career of petty crime of much the same genre ever since. How can such a man live with himself? Most people wonder. How can the psyche stand up to such cruel abuse?

In common with many of my colleagues, I do not find this crime bewildering and inexplicable; it strikes me as a straightforward felony of which scientists must be supposed no less capable than any other professional men. But what *is* surprising is to find crookedness of a kind that annuls everything that makes the scientific profession attractive, honorable, and praiseworthy.

There is no such person as *the* scientist, then, and *a fortiori* no such person as the wicked scientist, even though fiction of the kind in which a "Chinaman" could be relied upon to be the villain has now been supplanted by a fiction still lower-browed in which "the scientist" plays a rather similar role. Gothic fiction did not end with the writings of Mary Shelley and Mrs. Ann Radcliffe. In its modern equivalent, wicked scientists abound ("Soon ze whole vorld vill be in my power!" such a one cries out in a strong Central European accent). I feel that some of the fear laymen have of scientists is a judgment upon them for their passive acquiescence in the conventions of this childish literature.

I suppose it is possible that the stereotype of the wicked scientist dissuades some youngsters from entering the profession, but the world today is so topsy-turvy that perhaps as many are attracted as are repelled by the prospect of a career of malefaction.

The wicked scientist is no more implausible than that other

stereotype dating from the days of improving literature: the man with the dedicated and purposeful expression who, heedless of personal welfare or material reward, finds in the pursuit of truth a complete intellectual and spiritual diet. No, *scientists are people*—a literary discovery of C. P. Snow's; whatever the motives that persuade anyone to pursue a career in scientific research, a scientist must very much want to be one. In my anxiety that they should not be underestimated I may sometimes make too much of the vexations and frustrations of scientific life, but it can be one of great contentment and reward (I do not mean, though I do not exclude, material reward), with the added satisfaction of using one's energies to the full.

2

How Can I Tell If I Am Cut Out to Be a Scientific Research Worker?

People who believe themselves cut out for a scientific life are sometimes dismayed and depressed by, in Sir Francis Bacon's words, "The subtilty of nature, the secret recesses of truth, the obscurity of things, the difficulty of experiment, the implication of causes and the infirmity of man's discerning power, being men no longer excited, either out of desire or hope, to penetrate farther."

There is no certain way of telling in advance if the day-dreams of a life dedicated to the pursuit of truth will carry a novice through the frustration of seeing experiments fail and of making the dismaying discovery that some of one's favorite ideas are groundless.

Twice in my life I have spent two weary and scientifically profitless years seeking evidence to corroborate dearly loved hypotheses that later proved to be groundless; times such as these are hard for scientists—days of leaden gray skies bringing with them a miserable sense of oppression and inadequacy. It is my recollection of these bad times that accounts for the earnestness of my advice to young scientists that they should have more than one string to their bow and should be willing to take no for an answer if the evidence points that way.

It is especially important that no novice should be fooled by old-fashioned misrepresentations about what a scientific life is like. Whatever it may have been alleged to be, it is in reality exciting, rather passionate and—in terms of hours of work—a very demanding and sometimes exhausting occupation. It is also likely to be tough on a wife or husband and children who have to live with an obsession without the compensation of being possessed by it themselves (see "Hard Luck on Spouses?" in Chapter 5).

A novice must stick it out until he discovers whether the rewards and compensations of a scientific life are for him commensurate with the disappointments and the toil; but if once a scientist experiences the exhilaration of discovery and the satisfaction of carrying through a really tricky experiment—once he has felt that deeper and more expansive feeling Freud has called the "oceanic feeling" that is the reward for any real advancement of the understanding—then he is hooked and no other kind of life will do.

Motives

What about the motives for becoming a scientist in the first place? This is the kind of subject upon which psychologists might be expected to make some pronouncement. Love of finicky detail was said by Lou Andreas Salome to be one of the outward manifestations of—uh—"anal erotism," but scientists in general are not finicking, nor, luckily, do they often have to be. Conventional wisdom has always had it that curiosity is the mainspring of a scientist's work. This has always seemed an inadequate motive to me; *curiosity* is a nursery word. "Curiosity killed the cat" is an old nanny's saying, though it may have been that same curiosity which found a remedy for the cat on what might otherwise have been its deathbed.

Most able scientists I know have something for which "exploratory impulsion" is not too grand a description. Immanuel Kant spoke of a "restless endeavor" to get at the truth of things, though in the context of the not wholly convincing argument that nature would hardly have implanted such an ambition in our breasts if it had not been possible to gratify it. A strong sense of unease and dissatisfaction always goes with lack of comprehension. Laymen feel it, too; how otherwise can we account for the relief they feel when they learn that some odd and disturbing phenomenon can be explained? It cannot be the explanation itself that brings relief, for it may easily be too technical to be widely understood. It is not the knowledge itself, but the satisfaction of knowing that something is known. The writings of Francis Bacon and of Jan Amos Comenius—two of the philosophic founders of modern science whose writings I shall often refer to—are suffused by the imagery of light. Perhaps the restless unease I am writing of is an adult equivalent of that childish fear of the dark that can be dispelled, Bacon said, only by kindling a light in nature.

I am often asked, "What made *you* become a scientist?" But I can't stand far enough away from myself to give a really satisfactory answer, for I cannot distinctly remember a time when I did not think that a scientist was the most exciting possible thing to be. Certainly I had been stirred and persuaded by the writings of Jules Verne and H. G. Wells and also by the not necessarily posh encyclopedias that can come the way of lucky children who read incessantly and who are forever poring over books. Works of popular science helped, too: sixpenny—in effect, dime—books on stars, atoms, the earth, the oceans, and suchlike. I was literally afraid of the dark, too—and if my conjecture in the paragraph above is right, that may also have helped.

Am I Brainy Enough to Be a Scientist?

An anxiety that may trouble some novices, and perhaps particularly some women because of the socially engendered habit —not often enough corrected—of self-depreciation, is whether they have brains enough to do well in science. It is an anxiety they could well spare themselves, for one does not need to be terrifically brainy to be a good scientist. An antipathy or a total indifference to the life of the mind and an impatience of abstract ideas can be taken as contraindications, to be sure, but there is nothing in experimental science that calls for great feats of ratiocination or a preternatural gift for deductive reasoning. Common sense one cannot do without, and one would be the better for owning some of those old-fashioned virtues that seem unaccountably to have fallen into disrepute. I mean application, diligence, a sense of purpose, the power to concentrate, to persevere and not be cast down by adversity—by finding out after long and weary inquiry, for example, that a dearly loved hypothesis is in large measure mistaken.

An Intelligence Test. For full measure I interpolate an intelligence test, the performance of which will differentiate between common sense and the dizzily higher intellections that scientists are sometimes thought to be capable of or to need. To many eyes, some of the figures (particularly the holy ones) of El Greco's paintings seem unnaturally tall and thin. An ophthalmologist who shall be nameless surmised that they were drawn so because El Greco suffered a defect of vision that made him *see* people that way, and as he saw them, so he would necessarily draw them.

Can such an interpretation be valid? When putting this question, sometimes to quite large academic audiences, I have added, "Anyone who can see *instantly* that this explanation is nonsense and is nonsense for philosophic rather than aesthetic reasons is undoubtedly bright. On the other hand, anyone who still can't see it is nonsense even when its nonsensicality is explained must be rather dull." The explanation is epistemological—that is, it has to do with the theory of knowledge.

Suppose a painter's defect of vision was, as it might easily have been, diplopia—in effect, seeing everything double. If the ophthalmologist's explanation were right, then such a painter would paint his figures double; but if he did so, then when he came to inspect his handiwork, would he not see all the figures fourfold and maybe suspect that something was amiss? If a defect of vision is in question, the only figures that could seem natural (that is, representational) to the painter must seem natural to us also, even if we ourselves suffer defects of vision; if some of El Greco's figures seem unnaturally tall and thin, they appear so because this was El Greco's intention.

I do not wish to undervalue the importance of intellectual skills in science, but I would rather undervalue them than overrate them to a degree that might frighten recruits away. Different branches of science call for rather different abilities, anyway, but after deriding the idea that there is any such thing as *the* scientist, I must not speak of "science" as if it were a single species of activity. To collect and classify beetles requires abilities, talents and incentives quite different from, I do not say inferior to, those that enter into theoretical physics or statistical epidemiology. The pecking order within science—a most complicated *snobismus*—certainly rates theoretical physics above the taxonomy of beetles, perhaps because in the collection and classification of beetles the order of nature is thought to spare us any great feat of judgment or intellection: is not there a slot waiting for each beetle to fit into?

Any such supposition is merely inductive mythology, however, and an experienced taxonomist or paleobiologist will assure a beginner that taxonomy well done requires great deliberation, considerable powers of judgment and a flair for the discernment of affinities that can come only with experience and the will to acquire it.

At all events scientists do not often think of themselves as brilliantly brainy people—and some, at least, like to avow themselves rather stupid. This is a transparent affectation, though—unless some uneasy recognition of the truth tempts them to fish for reassurance. Certainly very many scientists are not intellectuals. I myself do not happen to know any who are Philistines unless—in a very special sense—it is being a Philistine to be so overawed by the judgments of literary and aesthetic critics as to take them far more seriously than they deserve.

Because so many experimental sciences call for the use of manipulative skills, it is part of conventional wisdom to declare that a predilection for or proficiency at mechanical or constructive play portends a special aptitude for experimental science. A taste for Baconian experimentation (see Chapter 9) is often thought significant, too—for example, an insistent inner impulsion to find out what happens when several ounces of a mixture of sulfur, saltpeter and finely powdered charcoal is ignited. We cannot tell if the successful prosecution of such an experiment genuinely portends a successful research career because only they become scientists who don't find out. To devise some means of ascertaining whether or not these conventional beliefs hold water is work for sociologists of science. I do not feel, though, that a novice need be turned away from science by clumsiness or an inability to mend radio sets or bikes. These skills are not instinctual; they can be learned, as dexterity can be. A trait surely incompatible with a scientific career is to regard manual work as undignified or inferior, or to believe that a scientist has achieved success only when he packs away test tubes and culture dishes, turns off the Bunsen burner, and sits at a desk dressed in collar and tie. Another scientifically disabling belief is to expect to be able to carry out experimental research by issuing instructions to lesser mortals who scurry hither and thither to do one's bidding. What is disabling about this belief is the failure to realize that experimentation is a form of thinking as well as a practical expression of thought.

Opting Out. The novice who tries his hand at research and finds himself indifferent to or bored by it should leave science without any sense of self-reproach or misdirection.

This is easy enough to say, but in practice the qualifications required of scientists are so specialized and time-consuming that they do not qualify him to take up any other occupation; this is especially a fault of the current English scheme of education and does not apply with the same force in America, whose experience of *general* university education is so much greater than our own.[1]

A scientist who pulls out may regret it all his life or he may feel liberated; if the latter, he probably did well to quit, but any regret he felt would be well-founded, for several scientists have told me with an air of delighted wonderment how very satisfactory it is that they should be paid—perhaps even adequately paid—for work that is so absorbing and deeply pleasurable as scientific research.

1. The great wave of university building that in England transformed city colleges into the civic universities happened about 1890–1910, but in America the mountain-building epoch of university evolution happened about a hundred years ago.

3

What Shall I Do Research On?

Old-fashioned scientists would say that anyone who was obliged to ask such a question had mistaken his métier, but this attitude dates from the time when a scientist newly graduated was believed to be equipped to embark upon research forthwith. It is far otherwise nowadays when apprenticeship is the almost invariable rule; today, a young hopeful attaches himself as a graduate student to some senior scientist and hopes to learn his trade and be rewarded by a master's degree or doctorate in philosophy as evidence that he has done so. (The Ph.D. is a passport valid for immigration into almost any academic institution in the world.) Even so, some choice should be exercised in choosing a patron in the first place and in deciding what to do after receiving a postgraduate degree.

I myself went through the motions of taking a D.Phil. at Oxford, was examined and duly given leave to pay the (in those days) quite large sum of money necessary to be registered as a D.Phil. and receive the appropriate accolade, but I decided not to do so. This proves that human life can persist without the doctor's degree (which was in any case highly unfashionable in the Oxford of my days—my own tutor, J. Z. Young, was not a doctor, though many honorary degrees have brought respectability to him since).

The easy way to choose a patron is to pick the person closest at hand—the head or other senior staff member in the department of graduation who may be on the lookout for disciples or an extra pair of hands. Such a choice will have the advantage that the graduate student need not change his opinions, lodgings, or friends, but conventional wisdom frowns upon it and is greatly opposed to young graduates' continuing in the same department; lips are pursed, the evils of academic inbreeding piously rehearsed, and sentiments hardly more lofty or original than that "travel broadens the mind" are urged upon any graduate with an inclination to stay put.

These abjurations should not be thought compelling. Inbreeding is often the way in which a great school of research is built up. If a graduate understands and is proud of the work going on in his department, he may do best to fall into step with people who know where they are going. A graduate student should by all means attach himself to a department doing work that has aroused his enthusiasm, admiration or respect; no good will come of merely going wherever a job offers, irrespective of the work in progress.

It can be said with complete confidence that any scientist of any age who *wants to make important discoveries must study important problems.* Dull or piffling problems yield dull or piffling answers. It is not enough that a problem should be "interesting"—almost any problem is interesting if it is studied in sufficient depth.

As an example of research work not worth doing, Lord Zuckerman invented the cruelly apt but not ridiculously far-fetched example of a young zoology graduate who has decided to try to find out why 36 percent of sea urchin eggs have a tiny little black spot on them. This is not an important problem; such a graduate student will be lucky if this work commands the attention or interest of anyone except perhaps the poor fellow next door who is trying to find out why 64 percent of sea urchin eggs do *not* have a little black spot on them. Such a student has committed a kind of scientific suicide, and his supervisors are very much to blame. The example is purely imaginary, of course, for Lord Zuckerman knows very well that no sea urchin eggs are spotted.

No, the problem must be such that it *matters* what the answer is—whether to science generally or to mankind. Scientists considered collectively are remarkably single-minded in their views about what is important and what is not. If a graduate student gives a seminar and no one comes or no one asks a question, it is very sad, but not so sad as the question gallantly put by a senior or a colleague that betrays that he hasn't listened to a word. But it is a warning sign, a shot across the bows.

Isolation is disagreeable and bad for graduate students. The need to avoid it is one of the best arguments for joining some intellectually bustling concern. It might be his own department, but if it is not, the graduate must resist all attempts by his seniors to persuade him to join it as a graduate student—a warning made necessary by the fact that some seniors are not above using a postgraduate stipendiary award within their gift as a bait to recruit students who would not otherwise have thought to come their way. In these days of disposable equipment, it has become too easy to treat a graduate student in the same spirit—as a disposable colleague.

After graduate students have taken their Ph.D.s, they must on *no account* continue with their Ph.D. work for the remainder of their lives, easy and tempting though it is to tie up loose ends and wander down attractive byways. Many successful scientists try their hands at a great many different things before they settle upon a main line of investigation, but this is a privilege that can be enjoyed only in the employment of very understanding seniors and when the graduate student has not been enlisted to do a particular job. If he has been, it is his duty to do it.

Because the newly graduated Ph.D. is still very much a beginner, a new migratory movement has grown up in modern science that is spreading as rapidly as the at one time newfangled habit (deplored in the Oxford of my days) of taking Ph.D.s at all. This new movement is the migration of "postdocs." Graduate research and attendance at conferences usually gives graduate students powers of judgment that they often wish they had had before they embarked on their graduate work. Later on they will know a great deal more than they did at first about the places where really exciting and important work is going on, preferably in congenial company. To one or other such group the most energetic postdocs will try to attach themselves. Senior scientists welcome them because as they have chosen to come they are likely to make good colleagues; for their part, the postdocs are introduced to a new little universe of research.

Whatever may be thought about the Ph.D. treadmill, this new postdoctoral revolution is an unqualifiedly good thing, and it is very much to be hoped that the patrons and benefactors of science will not allow it to languish.

In choosing topics for research and departments to enlist in, a young scientist must beware of following fashion. It is one thing to fall into step with a great concerted movement of thought such as molecular genetics or cellular immunology, but quite another merely to fall in with prevailing fashion for, say, some new histochemical procedure or technical gimmick.

4

How Can I Equip Myself to Be a Scientist or a Better One?

The number and complexity of the techniques and supporting disciplines used in research are so large that a novice may easily be frightened into postponing research in order to carry on with the process of "equipping himself." As there is no knowing in advance where a research enterprise may lead and what kinds of skills it will require as it unfolds, this process of "equipping oneself" has no predeterminable limits and is bad psychological policy, anyway; we always need to know and understand a great deal more than we do already and to master many more skills than we now possess. The great incentive to learning a new skill or supporting discipline is an urgent need to use it. For this reason, very many scientists (I certainly among them) do not learn new skills or master new disciplines until the pressure is upon them to do so; thereupon they can be mastered pretty quickly. It is the lack of this pressure on those who are forever "equipping themselves" and who show an ominous tendency to become "night-class habitués" that sometimes makes them tired and despondent in spite of all their diplomas and certificates of proficiency.

Reading. Very similar considerations apply to a novice's inclination to spend weeks or months "mastering the literature." Too much book learning may crab and confine the imagination, and endless poring over the research of others is sometimes psychologically a research substitute, much as reading romantic fiction may be a substitute for real-life romance. Scientists take very different views about "the literature"; some read very little, relying upon *viva voce* information, circulated "preprints," and the beating of tom-toms by which advances in science come to be known to those who want to know them. Such communications as these are for the privileged, though; they are enjoyed by those who have already made headway enough to hold views others would like to hear in return. The beginner *must* read, but intently and choosily and not too much. Few sights are sadder than that of a young research worker always to be seen hunched over journals in the library; by far the best way to become proficient in research is to get on with it—if need be, asking for help so insistently that in the long run it is easier for a colleague to help a novice than to think up excuses for not doing so.

It is psychologically most important to *get results,* even if they are not original. Getting results, even by repeating another's work, brings with it a great accession of self-confidence; the young scientist feels himself one of the club at last, can chip in at seminars and at scientific meetings with "My own experience was . . ." or "I got exactly the same results" or "I'd be inclined to agree that for this particular purpose medium 94 is definitely better than 93," and then can sit down again, tremulous but secretly exultant.

As they gain experience, scientists reach a stage when they look back upon their own beginnings in research and wonder how they had the temerity to embark upon it, considering how thoroughly ignorant and ill-equipped they were. That may well have been so; but fortunately their temperaments must have been sufficiently sanguine to assure them that they were not likely to fail where so many others not very unlike themselves had succeeded, and sufficiently realistic, too, to understand that their equipment would never be complete down to the last button—that there would always be gaps and shortcomings in their knowledge and that to be any good they would have to go on learning all their lives. I do not know any scientist of any age who does not exult in the opportunity continuously to learn.

Apparatus. Old-fashioned scientists sometimes insist on the disciplinary value of a scientist's making his own apparatus. If this is only a matter of piecing parts together, that is very well; but oscillographs, no. Most modern apparatus is far too sophisticated and complex to yield to do-it-yourself procedures; only under the very special circumstance that the equipment needed is not yet commercially available is it sensible to make it. Devising and constructing apparatus is a branch of the scientific profession; the novice should be content with one scientific career instead of trying to embark on two. He should not have time, anyway.

> Lord Norwich tried to mend the electric light
> It struck him dead—and serve him right!
> It is the duty of the wealthy man
> To give employment to the artisan.

It may not have been Lord Norwich, but it *was* Hilaire Belloc. Scientists are not wealthy, of course, but the scale of their grants is usually so adjusted as to make it possible to buy the equipment they need.

The Art of the Soluble. Following the lead of Bismarck and Cavour, who described the art of politics as "the art of the possible," I have described the art of research as "the art of the soluble."

By some people this was almost willfully misunderstood to mean that I advocated the study of easy problems yielding quick solutions—unlike my critics, who were studying problems of which the main attraction (to them) was that they could not be solved. What I meant *of course* was that the art of research is that of making a problem soluble by finding out ways of getting at it—soft underbellies and the like. Very often a solution turns on devising some means of quantifying phenomena or states that have hitherto been assessed in terms of "rather more," "rather less," or "a lot of," or—sturdiest workhorse of scientific literature—"marked" ("The injection elicited a marked reaction"). Quantification as such has no merit except insofar as it helps to solve problems. To quantify is not to be a scientist, but goodness, it does help.

My own career as a serious medical scientist began with devising a means for measuring the intensity of the reaction that a mouse or a man mounts against a graft transplanted upon it from some other mouse or some other man.

5

Sexism and Racism in Science

Women in Science

Throughout the world tens of thousands of women are engaged in scientific research or in science-based occupations—activities they are good at or bad at in much the same way and for much the same reasons as men are: they prosper who are energetic, intelligent, "dedicated" and hardworking, but languish who are lazy, unimaginative, or dull.

In view of the importance attached to "intuition" and insight in the chapter (11) devoted to the nature of the scientific process, we might—on the basis of the sexist illusion that women are especially intuitive in character—be tempted to expect women to be especially good at science. This view is not widely held by women, and I do not think it at all likely to be true because the "intuition" referred to above (that with which women are thought to be especially well endowed) connotes some special perceptiveness in human relations rather than the imaginative guesswork that is the generative act in science. But even if they are not *especially* proficient the scientific profession has special attractions for intelligent women; self-interest has long persuaded universities and the great research organizations to give women equality of treatment with men. This equality of treatment moreover is that which comes from equality of merit, not that which represents an enforced and perhaps reluctant acquiescence in the newly devised statutory obligations that now require employers to treat women as if they were human beings.

"It's fun being a woman scientist," such a one once said to me, "because you don't have to compete." Don't *have* to, maybe, but by golly you do, and are every bit as anxious about just acknowledgment of priority as the man next door, and just as obsessional and wrapped up in your work, too. It *is* fun being a scientist, that's for sure—though not for any reason that is thought to differentiate women from men.

Young women who enter a scientific profession and who may want to have children should examine their intended employers' rulings about maternity leave, time off with pay while enjoying it, and so on. The provision or nonprovision of day-care facilities is another consideration to bear in mind.

Young women anxious to defend the choice of a scientific career against the anxious and cautionary objections of parents and even old-fashioned school teachers should beware of citing Madame Curie as evidence that women can do well in science; any such tendency to generalize from isolated instances will convince no one that they have a natural aptitude for science —it is not Madame Curie but the tens of thousands of women gainfully and often happily engaged in scientific pursuits who should be called in evidence.

Although I have been the head of several laboratories in which women are employed, I have never been able to discern any distinctive style about their scientific work, nor have I any idea about how one would go about demonstrating any such distinction.

The case for rejoicing in the increasing number of women who enter the learned professions has nothing primarily to do with providing them with gainful employment or giving them an opportunity to develop their full potential. It is above all due to the fact that the world is now such a complicated and rapidly changing place that it cannot even be kept going (let alone improved, as we meliorists think it can be) without using the intelligence and skill of approximately 50 percent of the human race.

Hard Luck on Spouses? One of the scenes I remember most vividly from my period (1951–62) as professor (chairman) of zoology in University College London, the oldest and largest university in the federation that makes up "London University," was the gathering of teaching and research staff for coffee on Christmas morning.

What on earth were they doing there on Christmas Day? One or two were clearly lonely and had come to enjoy the special comradeship of travelers on the same road (the one that winds uphill all the way). Others came in to keep an eye on experiments in progress and incidentally to give the mice their Christmas dinner—the uproar created by a thousand mice eating corn flakes falls gratefully on the ears of those who are fond of mice and wish them well. But most of the men in the little gathering had it in common that they were fathers of young families. Back at the ranch, therefore, their wives were performing the daily miracle of young motherhood—entertaining, appeasing, suppressing the natural instincts of, and bringing out the best in, a family of children who seemed twice as numerous as they really were.

Men or women who go to the extreme length of marrying scientists should be clearly aware beforehand, instead of learning the hard way later, that their spouses are in the grip of a powerful obsession that is likely to take the first place in their lives outside the home, and probably inside too; there may not then be many romps on the floor with the children and the wife of a scientist may find herself disproportionately the man as well as the woman about the house when it comes to mending fuses, getting the car serviced, or organizing the family holiday. Conversely, the husband of a scientist must not expect to find *gigot de poulette cuit à la vapeur de Marjolaine* ready on the table when he gets home from work probably less taxing than his wife's.

Husband-and-Wife Teams. Some institutions make it a rule not to employ husband and wife in the same department, thus prohibiting the formation of married research teams. The rule was probably devised by tidy-minded administrators to prevent favoritism and the possibility of insufficiently "objective" appraisals of research. The rule is often thought wise because, through one of those tricks of selective memory I refer to elsewhere, we remember husband-and-wife teams that came apart more readily than those who got on well. There is room for research here by a competent sociologist of science, and until it is done, evaluation of the success of husband-and-wife research teams can only be surmise.

I should find it hard to believe that the conditions that must be satisfied if collaboration is to be successful (see Chapter 6) are any less exacting for married couples than for teams adventitiously formed.

I guess it to be a necessary condition for effective collaboration that husband and wife should love each other in the fully adult sense, that they should work together right from the beginning with that charity and mutual understanding which happily married couples may take many years to achieve.

Competition between man and wife is especially destructive, and although I thought at one time that there should not be too great an inequality of merit in husband-and-wife research teams, I am not now so sure. Things may be easier when competition is self-evidently fruitless.

It is an important point of manners, though, that members of married research teams should never attempt any public attribution of merit for the outcome of joint research—an attempt just as offensive when one partner allots all the credit to the other as when he takes it for him- or herself.

My reminder (in Chapter 6) that every member of a research team may have disagreeable personal habits that make collaboration more of a penance than a joy applies with equal force to married couples, though with the unhappy difference that the traditional candor of communication between man and wife may remove the mannerly embargo on telling a colleague how revolting he is; mannerliness does as much for collaboration as magnanimity, a principle that can hardly apply with lesser force to man and wife than to other teams.

Chauvinism and Racism More Generally

The idea that women are, and are to be expected to be, constitutionally different from men in scientific ability is a cozy domestic form of racism—of the more general belief that there are inborn constitutional differences in scientific prowess or capability.

Chauvinism. All nations like to think that there is something about them that makes them especially proficient in science. It is a source of national pride more elevated than the possession of a national airline or an atomic arsenal, or even prowess in football. *"La chimie, c'est une science française,"* said a contemporary of Lavoisier's, and I can still remember my schoolboy indignation at so presumptuous a claim. It is one that might much more justly have been made for German chemistry in the great days of Emil Fischer (1852–1919) and Fritz Haber (1868–1934), the days when young British and American chemists trooped over to Germany for their initiation into advanced biological chemistry and for one of those newfangled German Ph.D.s.[1]

Many Americans take it quite for granted that they are best in science, and sometimes enthusiastically quote evidence that they are so, of a kind that any trained sociologist could demolish out of hand. "Of course," I have heard it said in the bar of a suburban tennis club populated by young business executives, "the trouble with the Japanese is that they can only imitate others; they have no original ideas of their own." I wonder if the owner of that loud and confident voice—the very voice that at other times can be heard to declare that high speed in motor vehicles, so far from being a cause contributory to accidents, is actually conducive to safety—now realizes that the Japanese are inexhaustibly ingenious and inventive. The postwar flowering of Japanese science and science-based industry has already added great strength to science and technology throughout the world.

I know no nation in the running for such stakes that has not produced a number of highly able scientists and made a contribution to world science proportional to its size. Regional differences are intrinsically unlikely for methodological reasons, and no experienced scientist seriously believes that they exist. The jargon of nationalism is not part of the scientific vernacular. After a scientific lecture, no one ever hears it said: "Of course,

1. Nothing illustrates the importance attached to German chemistry more clearly than the fact that the study of German was for many years compulsory for would-be chemists.

half his slides were shown upside down, but that's a Serbo-Croat for you."

In those great research institutions that are concourses of all nations—the Institut Pasteur in Paris, the National Institute for Medical Research in London, the Max Planck Institute in Freiburg, the Institute for Cellular Pathology in Brussels, and Rockefeller University in New York—the nationality of the inmates is of little account and is seldom thought of. The numerical preponderance of Americans, their great generosity in funding research all over the world and in organizing conferences has brought it about that broken English is the international language of science. In international congresses the nations are distinguished not by styles of scientific research but by the emergence of different national styles in the delivery of scientific papers. The low, even monotone that is the nearest approach to an American national style contrasts amusingly with the rise and fall of the voice that Americans think so comic in an English delivery, and the English so comic in papers delivered by Swedes.

Intelligence and Nationality. I believe in the notion of "intelligence,"[2] and I believe also that there are inherited differences in intellectual ability, but I do not believe that intelligence is a simple scalar endowment that can be quantified by attaching a single figure to it—an I.Q. or the like.[3] Psychologists who do hold these opinions have been led into declarations so foolish as to make it hard to believe that they were not made up with the intention of bringing their subject into disrepute.

The application of "intelligence tests" to recruits into the U.S. forces in World War I, and even before then to would-be immigrants into the United States at the receiving station, Ellis Island, led to the compilation of a vast mass of intrinsically

2. I once spoke to a human geneticist who declared that the notion of intelligence was quite meaningless, so I tried calling him *un*intelligent. He was annoyed, and it did not appease him when I went on to ask how he came to attach such a clear meaning to the notion of lack of intelligence. We never spoke again.

3. P. B. Medawar, "Unnatural Science," *New York Review of Books* 24 (February 3, 1977), p. 13–18.

untrustworthy numerical information, the analysis of which led I.Q. psychologists into extremities of folly in which the following may never be surpassed: Henry Goddard's investigation into the intelligence of would-be immigrants led him to the conclusion that 83 percent of the Jews and 80 percent of the Hungarians seeking entry were feebleminded.[4]

Such a judgment upon Hungarians and Jews will be thought especially offensive by those who rightly or wrongly have come to believe that Jews have a special aptitude for science and the professions, and that such a constellation of talent as Thomas Balogh, Nicholas Kaldor, George Klein, Arthur Koestler, John von Neumann, Michael Polanyi, Albert Szent-Gyorgyi, Leo Szilard, Edward Teller and Eugene Wigner must point to something special about the Hungarian constitution.

But are not such opinions as offensively racist as those that are rightly the subject of public obloquy? No, they are not racist at all, for there is no implication here of genetic elitism; Hungarians are a political entity, not a race, and although the Jewish people have many of the biological characteristics of a race, there are many good nongenetic reasons why they should be especially good at science and scholarly activities generally: the traditional reverence of Jews for learning, the sacrifices Jewish families are prepared to make to give their children a start in one of the learned professions, the willingness of Jews to help each other, and the long and sad history that has convinced so many Jews that in a competitive and often hostile world the best hope of security and advancement lies in the learned professions.

As to the all-star cast of Hungarian intellectuals (many of them Jews, incidentally), any thought of a genetic interpretation may be instantly dispelled by the reflection that for this particular World Cup, a team of equal or even greater distinction could be recruited from Vienna or thereabouts: Herman Bondi, Sigmund Freud, Karl von Frisch, Ernst Gombrich, F. A. von Hayek, Konrad Lorentz, Lisa Meitner, Gustav Nossall, Max

---

4. L. J. Kamin, *The Science and Politics of IQ* (New York: John Wiley & Sons, 1974), p. 16. Goddard's views are from the *Journal of Psycho-Asthenics (sic)* for 1913.

Perutz, Karl Popper, Erwin Schrödinger and Ludwig Wittgenstein.

The cosmogeny of these remarkable constellations of talent is something for the historian of culture and the historian-sociologist to ponder upon and interpret.

If, as I believe, scientific inquiry is an enormous potentiation of common sense, then the absence of any important national differences in the ability to "do" science may be thought to uphold Descartes's contention that common sense is the most equitably distributed of all human gifts.

6

Aspects of Scientific Life and Manners

A scientist soon discovers that he has become a member of the cast of *them* in the context "What mischief are *they* up to now?" or "*They* say we shall colonize the Moon in fifty years."

Scientists naturally want to be thought well of and, like other professional men, would like their calling to be respected. They will find from the beginning, however, that upon learning that they are scientists, the people they meet tend to adopt one or another of two opinions, which cannot both be right: that because a man is a scientist his judgment on any topic whatsoever is either (a) specially valuable or (b) virtually worthless. These opinions are of that habitual and inflexible kind which we tend to associate with political beliefs and are every bit as difficult to reason with or alter. An attempt should nevertheless be made not to acerbate either condition of mind. "Just because I am a scientist doesn't mean I'm anything of an expert on . . ." is a formula for all seasons; the sentence may be completed in almost as many different ways as there are different topics of conversation. Proportional representation, the Dead Sea Scrolls, the fitness of women for holy orders, or the administrative problems of the eastern provinces of the Roman Empire are examples enough, but when the subject is carbon dating or the likelihood of there being constructed a machine of perpetual motion, a scientist may allow himself the benefit of a few extra decibels to give his voice something of a cutting edge.

The cruel presumption of his Philistinism may sometimes prompt a scientist to pretend to cultural interests and a cultural understanding he does not really possess; in extreme cases, his audience may have to put up with a little parade of secondhand cultural aperçus from fashionable critics or imperfectly remembered excerpts from the *Meditations of Cardinal Poggi Bonsi.*

Scientists should be on their guard, though. Humbug is usually easy to identify, and in scientists easier than most, for if they are not used to intellectual or literary chatter, they are all too likely to give themselves away by mispronunciations that no one will correct or by cultural misconceptions so vast that no one will think it worthwhile to dispute them.

Cultural Revenge. A scientist who has been culturally snubbed or who feels himself otherwise at a disadvantage may sometimes solace himself by a sour withdrawal from the world of the humanities and the fine arts. An alternative medication for bruised psyches is to become a Knowall; one's audience is thenceforward bedazzled by fashionable talk of scenarios, paradigms, Gödel's theorem, the import of Chomsky's linguistics and the extent of Rosicrucian influences on the fine arts. A savage revenge indeed, but one that will make a scientist's former companions flee in disorder on his arrival. No form of words is more characteristic of the Knowall than the following: "Of course, there is really no such thing as x; what most people call x is really y." In this context, x can be almost anything people believe in, such as the Renaissance, the Romantic revival or the Industrial Revolution; y is usually something declared to be stirring for the first time in the bosoms of the proletariat. Becoming a Knowall is not, however a serious occupational risk of scientists; the worst Knowalls I have known were both economists.

Whichever form of revenge a scientist decides upon, whether to withdraw from cultural interests or to dazzle his fellow men with omniscience, he should certainly ask himself, "Whom am I punishing?"

Cultural Barbarism and the History of Science. Scientists are assumed to be illiterate and to have coarse or vulgar aesthetic sensibilities until the contrary is proved; however much it may annoy, a young scientist must again be warned against attempting any parade of culture to rebut this imputation. In any case, the accusation is in one respect well-founded: I have in mind the total indifference of many young scientists to the history of ideas, even of the ideas that lie at the root of their own research. I have tried in *The Hope of Progress* to excuse this attitude of mind, pointing out that the growth of science is of a special kind and that science does in some sense contain its cultural history within itself; everything that a scientist does is a function of what others have done before him: the past is embodied in every new conception and even in the possibility of its being conceived at all.

A most distinguished French historian, Fernand Braudel, has said of history that "it devours the present." I do not quite understand what he means (those profound French epigrams, you know), but in science, to be sure, it is the other way about: the present devours the past. This does something to extenuate a scientist's misguided indifference to the history of ideas.

If it were possible to quantify knowledge or degree of understanding and plot it as a graph against time, it would not be so much the height of the curve above the baseline as the total area accumulated between the two that would most faithfully represent the state of science at any one time.

However that may be, an indifference to the history of ideas *is* widely interpreted as a sign of cultural barbarism—and rightly, too, I should say, because a person who is not interested in the growth and flux of ideas is probably not interested in the life of the mind. A young scientist working in an advancing field of research should certainly try to identify the origin and growth of current opinions. Although self-interest should not be his motive, he will probably end with a stronger sense of personal identity if he can see where he fits into the scheme of things.

Science and Religion. "His is a gentleman's religion," the dialogue goes.

"And pray, sir, what is that?"

"Gentlemen do not discuss religion."

I have always thought this an exceptionally disagreeable fragment of dialogue, which reflects credit on no one. If for "gentleman" one substitutes "scientist," the story is in no way improved, but it becomes a more genuine description of very many scientists' lack of religious conviction.

There is no quicker way for a scientist to bring discredit upon himself and on his profession than roundly to declare—particularly when no declaration of any kind is called for—that science knows or soon will know the answers to all questions worth asking, and that the questions that do not admit a scientific answer are in some way nonquestions or "pseudoquestions" that only simpletons ask and only the gullible profess to be able to answer.

I am happy to say that however many scientists may think this, very few nowadays are mugs enough or rude enough to say so in public. Philosophically sophisticated people know that a "scientific" attack upon religious belief is usually no less faulty than a defense of it. Scientists do not speak on religion from a privileged position except insofar as those with a predilection for the Argument from Design have better opportunities than laymen to see the grandeur of the natural order of things, whatever they may make of it.

When Science Is to Be Defended

I hope I shall not be thought to be urging a humble demeanor upon scientists generally, but for the sake of the profession they should take pains not to bring it into disrepute; it is no longer taken for granted that science and civilization stand shoulder to shoulder in a common endeavor to work for the betterment of mankind. Scientists will certainly encounter and must work out some suitable means for rebutting the notion that, so far from trying to better the lot of mankind, the outcome of their work is to devalue much of what ordinary folk hold dear. Through science, you may hear, art has been replaced by artifice: portraiture by photography, live music by Muzak, good food by processed substitutes, and the old-fashioned crusty loaf by a chemically bleached or otherwise "improved," devitaminized, revitaminized, steam-baked, presliced parallelepiped in a polyethylene shroud.

This is an old story, though, which has more to do with avarice, the convenience of manufacturers and dishonest dealing than with science. Early in the nineteenth century William Cobbett, believing that all working people should bake their own bread, scathingly denounced a shop bread that we ourselves should probably have thought delicious, accusing it of being balderdashed with alum, filled out with potato flour, and having no more of the "natural sweetness of grain than is to be found in the offal which comes from the sawing of deal boards."

It is not really an adequate defense of modern "food science" to say that the reason why the stuff is made is that people want to buy it; such a defense disregards the well-known economic principle that supply creates demand, particularly if the supply is accompanied by meretricious advertisements creating the impression that a presliced bread substitute is in reality more natural and more deeply suffused by the sunshine of a cornfield than that which we used to buy at the little corner bakery before it was pulled down to make way for the supermarket. But be fair to science; it was scientists themselves who demonstrated that bread made from whole natural grains and unpolished rice is much better for us than polished white rice or that bleached, devitaminized, revitaminized . . . and so on. But it is no use expecting people to applaud a remedy for a disease they need never have had.

Is Science Undervalued?

Scientists sometimes feel a little aggrieved that most ordinary folk are so little interested and impressed by their calling.

The explanation of this real or seeming indifference was agreed upon by Voltaire and Samuel Johnson—a conjunction of opinion so unlikely that there must surely be something in it. The explanation is true, so scientists had best come to terms with it, resent it though they may. Science does not have a major bearing on human relationships: on the relationship of governors to the governed; on *les passions de l'âme;* nor on the causes of exaltation or misery and the character and intensity of aesthetic pleasures.

In his *Dictionnaire philosophique,* Voltaire said that natural science "is so little essential for the conduct of life that philosophers didn't need it; it required centuries to learn a part of the laws of nature, but a day was enough for a sage to learn the duties of man."

In his *Life of Milton,* Dr. Samuel Johnson, chiding Milton and Abraham Cowley for entertaining the idea of an academy in which the scholars should learn astronomy, physics and chemistry in addition to the common run of school subjects, wrote:

> the truth is that the knowledge of external nature and of the sciences which that knowledge requires or includes, is not the great or the frequent business of the human mind. Whether we provide for action or conversation, whether we wish to be useful or pleasing, the first requisite is the religious and moral knowledge of right and wrong; the next is an acquaintance with the history of mankind, and with those examples which may be said to embody truth, and prove by events the reasonableness of opinions. Prudence and justice are virtues and excellencies of all times and of all places; we are perpetually moralists, but we are geometricians only by chance. Our intercourse with intellectual nature is necessary; our speculations upon matter are voluntary, and at leisure. Physical knowledge is of such rare emergence, that one man may know another half his life without being able to estimate his skill in hydrostatics or astronomy; but his moral and prudential character immediately appears.

There is no reason why these truths should diminish a scientist's self-esteem or lessen his contentment—even exultation—at being a scientist. Scientists whose work is prospering and who find themselves deeply absorbed in and transported by their research feel quite sorry for those who do not share the same sense of delight; many artists feel the same, and it makes them indifferent to—and is certainly a fully adequate compensation for—any respect they think owed them by the general public.

Collaboration

As nearly all my scientific work has been done in collaboration with others, I regard myself as an authority on the subject.

Scientific collaboration is not at all like cooks elbowing each other from the pot of broth; nor is it like artists working on the same canvas, or engineers working out how to start a tunnel simultaneously from both sides of a mountain in such a way that the contractors do not miss each other in the middle and emerge independently at opposite ends.

It is, in the planning stage, anyway, more like a session of gag writers, for although each one knows, as all scientists know, that having an idea—a brainwave—can be only a personal event, each also knows that an atmosphere can be created in which one member of the team sparks off the others so that they all build upon and develop each other's ideas. In the outcome, nobody is quite sure who thought of what. The main thing is that something was thought of. A young scientist who feels a strong compulsion to say "That was my idea, you know," or "Now that you have all come round to my way of thinking . . ." is not cut out for collaborative work, and he and his colleagues would do better if he worked on his own. Old hands will always congratulate a beginner on a bright idea that was genuinely the beginner's and not a product of the synergism of minds that such a session promotes. *Synergism* is the key word in collaboration—it connotes that the joint effort is greater than the sum of the several contributions to it—but collaboration is not obligatory, no matter how many pompous pronouncements may be made on the supersession of the individual by the team. Collaboration is a joy when it works, but many scientists can and many do get on very well as loners.

A few Polonian precepts can do something to indicate whether or not a scientist is cut out for collaboration. Unless he likes his colleagues and admires them for their special gifts, he should shun it; collaboration requires some generosity of spirit, and a young scientist who can recognize in himself an envious temperament and is jealous of his mates should on no account try to work with others.

Each teammate should intone to himself from time to time, "Amazing though it may seem, I, too, have a number of behavioral traits that make it almost miraculous that anyone can put up with me: my slowness over figures, for example, my practice of whistling vocal gems from the operas through that gap in my teeth, and my habit of losing crucial documents (such as the only key to the double-blind trial)."

"My own faults as a colleague?" did someone ask? I thought someone would bring that up. Grave and numerous, surely, but not to such a degree as to have lost me the friendship of anyone I have ever worked with. I especially enjoy collaboration and have been rewarded by benefiting all my life from the collaboration of a succession of unusually able and likable colleagues.

When the time comes for the collaborative work to be published, a young scientist will naturally expect to figure in the credit titles, but not more prominently than his colleagues think fair—they will not do him down. I myself like and have usually adopted the Royal Society's alphabetical rule, believing that the rebuffs and disappointments of the world's Zygysmondis are in the long view counterbalanced by the undeservedly good fortune enjoyed by the Aaronsons of the world.

Technicians as Colleagues. When I began research, it was taken for granted at Lord's Cricket Ground, the game's headquarters, that so deep a cultural and social abyss separated professionals from amateurs that they should enter the playing field by different gates, even when members of the same team; at Wimbledon, professionals were not even allowed to compete. There is more sense in the latter ruling, for at lawn tennis amateurs need protection from professionals, whereas cricket, as George Orwell pointed out, has the remarkable distinction that amateurs can hold their own against pros.

Something of the same *snobismus* was at that time extended as a matter of course to technicians, who were regarded as laboratory servants to fetch and carry, do most of the more tedious or smellier jobs, and execute faithfully the instructions of the maestro who sat at his desk having great thoughts. This has all changed—and very much for the better. Technicians' jobs are now sought after to a degree that makes it possible for employers to insist upon entry standards as high as those that admit to universities. With a recognized career structure and increasing confidence in their own abilities, technicians have gone up in their own estimation, too—a most important element in "job satisfaction." Technicians often are, and always ought to be, better than "academic" or teaching staff at certain theoretical or practical operations. "Ought to be" because a technician can sometimes be more specialized than the staff member whom he helps: teaching duties or administration and a variety of other commitments may often oblige an academic staff member to keep more balls in the air than a technician, and he may have too many graduate or undergraduate students to make it possible for him to become adequately proficient at all the things he should be able to do.

Although such a declaration will shock the diehards who still live in the days when it was thought proper for professionals to be excluded from the courts, technicians are colleagues in a collaborative research; they must be kept fully in the picture about what an experiment is intended to evaluate and about the way in which the procedures decided upon by mutual consultation might "conduce to the sum of the business" (Bacon).

Technicians who have enough good sense to prosper in their jobs soon learn how to impress upon young scientists that despite their paper degrees and *cum laudes* they still have a lot to learn about scientific research—and no lesson sooner than to treat their technicians as fellow workers. For their part (see the section on "The Truth," below), technicians must always shun any tendency to tell the people whom they help the results they would most like to hear, as Mendel's gardeners may have done; though it is to be hoped that relations between them are not so bad as to give a technician any pleasure in being the bearer of unwelcome news.

Collaboration can lead to lifelong friendships or enmities. To the former, surely, if the partners to it are—in the coarse patois of my laboratory—magnanimice. If so, collaboration should be a joy, and if it is not, it must be brought to an end with the least possible delay.

Moral and Contractual Obligations

A scientist will normally have contractual obligations to his employer and has always a special and unconditionally binding obligation to the truth.

There is nothing about being a scientist that diminishes his obligation to obey the Official Secrets Act and the company's rules on not chatting confidingly about manufacturing procedures to bearded strangers in dark glasses. Equally, though, there is nothing about being a scientist that should or need deafen him or close his mind to the entreaties of conscience.

Contractual obligations on the one hand, and the desire to do what is right on the other, can pose genuinely distressing problems that many scientists have to grapple with. The time to grapple is *before* a moral dilemma arises. If a scientist has reason to believe that a research enterprise cannot but promote the discovery of a nastier or more expeditious quietus for mankind, then he must not enter upon it—unless he is in favor of such a course of action. It is hardly possible that a scientist should recognize his abhorrence of such an ambition the first time he stirs the caldron. If he *does* enter upon morally questionable research and then publicly deplores it, his beating of the breast will have a hollow and unconvincing sound.

The Truth

Any scientist who is reasonably inventive and imaginative is certain to make mistakes over matters of interpretation; certain, that is to say, to take a wrong view or propound a hypothesis that does not stand up to criticism. If that is all the mistake amounts to, no great harm is done and no sleep need be lost. It is an ordinary part of the hurly-burly of scientific life; it is not so serious, for where one guesses wrong, another may guess right. If, on the other hand, the mistake is over a matter of fact—if the scientist said the litmus paper turned red when in reality it turned blue—then there is indeed good reason to lose sleep and to be tormented by those cruel early-morning thoughts in which one sees oneself discredited, for such a mistake as this might make it very difficult or even impossible for someone else to interpret a scientist's findings aright—that is, to hit upon a hypothesis that will accommodate them.

I still vividly remember a most miserable time during which I believed that I had committed and sent to press a really serious factual blunder about the existence in white guinea pigs' skin of a nonpigmentary analogue of cells that in colored animals would have been manufacturing pigments. I still remember, too, my gratitude to a young colleague who went over the ground again with great care and set my mind at rest. This reassuring action depended upon the use of a microanatomical technique which required that a certain treatment should be administered to a tissue for twenty-four hours. I urged him to cut corners and shorten the treatment, but the naval discipline of his service days made him stick to the letter of the instructions; we waited twenty-four hours, during which I was miserably drafting letters of recantation to *Nature.* It is a lucky scientist who never has such bad moments.

This is oversimplified, of course; it assumes—as all scientists tend to assume—that there is a clear and easily recognizable distinction between fact and theory, between the information delivered by the senses and the construction that is put upon it. No modern psychologist would take such a view, nor did William Whewell when he pointed out that even that which seems to be the simplest sensory apprehension depends upon an act of mind for its interpretation: "There is a mask of theory over the whole face of Nature."[1]

Mistakes. If in spite of the most anxious precautions a scientist makes a mistake about a matter of fact—if the results were due to an impurity in a supposedly pure enzyme preparation or because hybrid mice were used in error for mice of an inbred strain—then the mistake must be admitted with the least possible delay. Human nature is such that the scientist may even gain credit from such a declaration and will not lose face—except perhaps in the bathroom mirror.

The important thing is not to try to lay down some voluminous smoke screen to conceal a blunder. I once knew an able scientist who claimed that cancer cells that had been frozen and dried in the frozen state could still propagate a tumor. The claim was mistaken, for the tissues that he thought were dry—though they looked so (and we had the author's word for it that they could be blown around the room)—still contained about 25 percent of moisture. Instead of recanting, the poor fellow rather damaged his subsequent research career by the pretense that the phenomenon he was really studying was the biophysics

1. William Whewell, *The Philosophy of the Inductive Sciences,* 2d ed. (London, 1847), pp. 37–42.

of cellular freezing itself and not the property that was thought to survive it. If he had admitted his mistake and got on with something else, he could have made a worthwhile contribution to science.

Though faulty hypotheses are excusable on the grounds that they will be superseded in due course by acceptable ones, they can do grave harm to those who hold them because scientists who fall deeply in love with their hypotheses are proportionately unwilling to take no as an experimental answer. Sometimes instead of exposing a hypothesis to a cruelly critical test (*to il cimento,* see Chapter 9), they caper around it, testing only subsidiary implications, or follow up sidelines that have an indirect bearing on the hypothesis without exposing it to the risk of refutation. I have witnessed this very procedure in a Russian laboratory whose existence depended on the efficacy of a serum that in the opinion of most foreigners simply did not have the properties claimed for it.

I cannot give any scientist of any age better advice than this: the intensity of the conviction that a hypothesis is true has no bearing on whether it is true or not. The importance of the strength of our conviction is only to provide a proportionately strong incentive to find out if the hypothesis will stand up to critical evaluation.

Poets and musicians may easily think this sadly cautionary advice and characteristic of the spiritless fact-finding that they suppose scientific inquiry to be. For them, I guess, what is done in a blaze of inspiration has a special authenticity. I guess also that this is true only where there is talent bordering upon genius.

A scientist who habitually deceives himself is well on the way toward deceiving others. Polonius foresaw it clearly ("This above all . . .").

Life-style

Although I firmly believe that creativity in the domain of scientific ideas is cognate with creativity as it occurs in poets, artists, and the like, the kinds of conventional wisdom or romantic nonsense that have grown up about circumstances conducive to one or other form of creativity differ in a number of ways.

To be creative, scientists need libraries and laboratories and the company of other scientists; certainly a quiet and untroubled life is a help. A scientist's work is in no way deepened or made more cogent by privation, anxiety, distress, or emotional harassment. To be sure, the private lives of scientists may be strangely and even comically mixed up, but not in ways that have any special bearing on the nature and quality of their work. If a scientist were to cut off an ear, no one would interpret such an action as evidence of an unhappy torment of creativity; nor will a scientist be excused any *bizarrerie,* however extravagant, on the grounds that he is a scientist, however brilliant. Ronald Clarke, writing on the life of J. B. S. Haldane,[2] described how his marital irregularities attracted the attention of the Sex Viri, a sort of buffo male voice sextet who watched over Cambridge's moral welfare and tried to deprive Haldane of his readership (the English equivalent of an associate professorship) on the grounds of immorality. The scenes accompanying the divorce that freed Charlotte Burghes to become Haldane's first wife do indeed read like the libretto of comic opera.

A scientist's or other research worker's need for tranquillity makes him sound dreadfully dull and pitifully unlike the stereotype of the creative artist of nineteenth-century romantic fiction—*la vie de Bohème* and all that.

Secure in their knowledge that research provides for a deeply absorbing and intellectually passionate life, scientists are not put out by, though they may wonder at, William Blake's "coming in the grandeur of inspiration to cast off rational demonstration," and with it, Bacon, Locke, and Newton.

The stereotype that represents "the scientist" as someone coldly engaged in the collection of facts and in calculations based upon them is no less a caricature than that which makes a poet poor, dirty, disheveled, tubercular maybe, and periodically in the grip of a poetic frenzy.

---

2. *The Life and Work of J. B. S. Haldane* (London: Hodder and Stoughton, 1968). See especially pp. 75–77.

Priority

Those who are anxious to discredit scientists, and especially the notion (not held by scientists themselves) that they are engaged in a cool, lofty and dispassionate quest for truth, are fond of calling attention to their anxiety about matters to do with priority—an anxiety that the work or ideas that the scientist believes to be his own should be credited to him and not to any others.

This anxiety is sometimes thought to be a new one—to be a natural consequence of the obligation upon a modern scientist to hold his own in a crowded and competitive world—but it is indeed *not* new; the research of Dr. Robert K. Merton[3] and his school has made it entirely clear that disputes over priority, sometimes of an especially venomous and unforgiving kind, are as old as science itself. It is a natural consequence of the fact that when several scientists are trying to solve the same problem, more than one may hit upon a solution—or *the* solution if there is only one.

When there *is* a unique solution—as, for example, the crystalline structure of DNA—the pressure is especially severe. Artists, I suspect, are a little contemptuous of a scientist's anxiety for credit, but then their situation is in no way comparable. If more than one poet or musician were to be invited to compose a patriotic ode or celebratory fanfare, the author of either would be furious if his own ode or fanfare were to become public as the work of another. But the problems that confront them do not have a unique solution; that two poets should hit upon the same wording or two composers the same score for

3. R. K. Merton, "Behavior Patterns of Scientists," *American Scientist* 57 (1969): 1–23. See also R. K. Merton, "Priorities in Scientific Discovery," *American Sociological Review* 22 (December 1957): 635–59; "Singletons and Multiples in Scientific Discovery," *Proceedings of the American Philosophical Society* 105 (October 1961): 470–86; "The Ambivalence of Scientists," *Bulletin of the Johns Hopkins Hospital* 112 (1963): 77–97; "Resistance to the Systematic Study of Multiple Discoveries in Science," *European Journal of Sociology* 4 (1963): 237–82; *On the Shoulders of Giants* (New York: The Free Press, 1965; Harcourt, Brace and World, 1967); "The Matthew Effect in Science," *Science* 159 (January 5, 1968): 56–63.

their respective acts of homage is statistically inconceivable, and—as I have pointed out on another occasion—the twenty years Wagner spent on composing the first three operas of *The Ring* were not clouded by the fear that someone else might nip in ahead of him with *Götterdämmerung.*

Whenever pride of possession is an important consideration —especially when the property in dispute is an idea—most people feel a strong sense of ownership. The investigative journalist with his special story or insight, the philosopher or historian with his mind-clearing way of looking at things, the administrator who hits upon just that disposition of funds or responsibility which will get around a tricky or confusing situation—each one feels that if the idea was his, it should be acknowledged to be so. Indeed, anxiety about priority is to be found in all walks of life, I believe. Sometimes, as with car or dress designers, it is a matter of securing their livelihood, but sometimes it is an aggressive arrogance; Field Marshal Lord Montgomery of Alamein, I have learned, was rapacious in his hunger for personal credit, sometimes when it was not deserved.

Problems to do with priority are especially acute in science because scientific ideas must eventually become public property, so that the only sense of ownership a scientist can ever enjoy is that of having been the *first* to have an idea—to have hit upon a solution or *the* solution before anyone else. I see nothing wrong in pride of possession, though in a scientific context, as in any other, possessiveness, meanness, secretiveness and selfishness deserve all the contempt they get. A lofty attitude to a scientist's pride of possession shows a sad lack of human understanding.

Secretiveness in a scientist is a disfigurement, to be sure, but it has its comic side; one of the most comically endearing traits of a young research worker is the illusion that everyone else is eager to hurry off to do his research before he can. In reality, his colleagues want to do their own research, not his. A scientist who is too cagey or suspicious to tell his colleagues anything will soon find that he himself learns nothing in return. G. F. Kettering, the well-known inventor (antiknock gasoline additives) and cofounder of General Motors, is said to have remarked that anyone who shuts his door keeps out more than he lets out. The agreed house rule of the little group of close colleagues I have always worked with has always been "Tell everyone everything you know"; and I don't know anyone who came to any harm by falling in with it. It is a good rule because a scientist's own work is so compellingly interesting and important that he is giving a colleague a great treat by telling all. But such a scientist must play fair; if he tells his colleagues all about his work, he must compose his mind to be spellbound by theirs, too. Of all the little episodes of the human comedy one may hope to witness in a research laboratory, none is more diverting than the scene in a corridor in which a young scientist (with a glittering eye, mayhap—and bearded, as like as not) stoppeth one (and possibly as many as three) of his colleagues, and tries to tell the whole story from beginning to end.

After circling around loftily in a holding pattern, discussions of priority usually end by discussing James D. Watson and *The Double Helix,* in which we see the hunger for priority in its most acute form. I defended Watson in *The Hope of Progress* for exactly the reasons that have prompted me to take an excusatory attitude toward the desire for recognition. Before passing judgment on Watson himself, literary folk should reflect that writers tend to be excused almost any behavior, however disagreeable or bizarre, if their work reveals in them an authentic genius. Jim Watson was a very bright young man indeed, and I have no hesitation in saying that *The Double Helix* is a classic. It is a matter therefore for regret rather than censure that in innumerable ways—especially in his failure to give credit where it belonged—the young Watson showed himself not big enough by far to match the truly splendid discovery in which he played so important a part.

Scientmanship. The word formation is of course Stephen Potter's: scientmanship is one-upmanship in a scientific context. As to *scientman,* Onions in his etymological dictionary[4] cites it as one of the solutions to the problem of what one word might

4. C. T. Onions, ed., *The Oxford Dictionary of English Etymology* (Oxford: Clarendon Press, 1966).

describe a man of science. Whewell did not coin the word *scientist* until 1840. Whewell was far and away science's greatest nomenclator. One of the publications of the Royal Society records the correspondence between Whewell and Michael Faraday about what names to attach to the opposite poles of the electrolytic cell. Faraday had come up with voltaode and galvanode, dexiode and skiaode, eastode and westode, zincode and platinode. There is a conscious air of finality about Whewell's "My dear Sir . . . I am disposed to recommend . . . *anode* and *cathode.*" And such they have been ever since.

Scientmanship comprehends the techniques used in the hope of enlarging one's reputation as a scientist or diminishing the reputation of others by nonscientific means. The practices of scientmanship are wholly discreditable and sadly betray a total absence of magnanimity. It is an old story, though: R. K. Merton has recalled Galileo's feeling of grievance at a rival who "tried to diminish whatever praise belonged to him for the invention of the telescope for use in astronomy."

This is an especially mean-minded form of scientmanship; a scientist who has picked up someone else's ideas may go to some lengths to create the impression that both he and the scientist to whom he is indebted derived the idea independently from some much older source. I can remember being surprised and hurt by the lengths to which a former friend—using the technique I have just mentioned—went to avoid acknowledging his indebtedness to me for the motivating idea of his research.

Another dirty trick is to cite only the most recent of a long string of scientific papers written by authors to whom you are indebted, while citations of your own research go back for years and years. It is a discreditable—indeed, an unforgivable—trick of scientmanship that withholds from a published paper some details of technique to prevent someone else from taking up the story where its author left off or alternatively to prevent someone else from proving that his story is pure science fiction. People who use such tricks probably think less of themselves for doing so, and this opinion is shared by everyone in the know—in fact, by many whose good opinion the culprit would most welcome.

Another trick of those who practice scientmanship is to

affect the possession of a mind so finely critical that no evidence is ever quite good enough ("I am not very happy about . . ."; "I must say I am not at all convinced by . . ."). Another such trick is to imply that one has thought of it or done it all before ("That is exactly what I thought when I got similar results in Pasadena"). I once knew a senior medical scientist who was so scathingly critical of everyone else's research that one wondered if his constitution precluded any possibility of belief. Considering his intelligence, he was expectedly barren of ideas of his own (something that may help to explain his critical temperament), but when he *did* have an idea of his own—my goodness, was it not the most important and profoundly illuminating notion that the world had ever known. On this topic all his critical faculties were suspended: he was a complete sucker for his own idea; any opposition to it aroused a degree of resentment that fell not far short of active enmity.

Scientists know very well when they get up to such tricks as these, and I suspect that each time they are used they bring with them a sense of inadequacy and self-diminishment. This is a pity, for his own good opinion is not the least important of those a scientist seeks to win.

The Snobismus *of Pure and Applied Science*

One of the most damaging forms of snobbism in science is that which draws a class distinction between pure and applied science. It is perhaps at its worst in England, where the genteel have a long history of repugnance to trade or any activity that might promote it.

Such a class distinction is particularly offensive because it is based upon a complete misconception of the original meaning of the word *pure*—the meaning that was thought to confer a loftier status upon pure than upon applied science. The word was originally used to distinguish a science of which the axioms or first principles were known not through observation or experiment—vulgar activities both—but through pure intuition, revelation, or a certain quality of self-evidence. Secure in his privileged access to the Absolute, the pure scientist felt one up on a man who dissected dead animals, calcined metals, or mixed chemicals to bring about various improbable conjunctions of natural events. All such activities seemed to scholars—and still did to my humanist colleagues when I was a young teacher at Oxford—to be rather inferior and infra dig, and to savor altogether too much of the tradesman or artisan; the applied scientist was unfit for the drawing room, and in spite of efforts to be broad-minded, even the least fastidious would be put off ("How would you feel if *your* sister wanted to marry an applied scientist?"). Had not my Lord Bacon identified pure science with *light*—with the kindling of a light in nature?—and had not God thought fit to create light before he turned his thoughts to applied science?

This snobbishness has lasted more than three hundred years; an historian of the Royal Society wrote thus in 1667 (the "inventions" he refers to are artificial devices and contrivances—that is, the arts). (In the context of the Royal Society toast to "The Arts and Sciences" or of the Royal Society of Arts [nothing to do with the Royal Society of London for improving natural knowledge], the "arts" are crafts, devices, and contrivances—that is, the various means by which thought is embodied in or translated into action.)

> Invention is an heroic thing and placed above the reach of a low and vulgar genius. It requires an active, a bold, a nimble, a restless mind: a thousand difficulties must be contemned with which a mean heart would be broken: many attempts must be made to no purpose: much treasure must be scattered without any return: much violence and vigor of thought must attend it: some irregularities and excesses must be granted that could hardly be pardoned by the severe rules of prudence.[5]

But Thomas Sprat was not one to believe that applied science could get on without a background of experimental philosophy: "The surest increase remaining to be made in the manual arts, is to be performed by the conduct of experimental philosophy. . . . Power rests on knowledge."[6] It may strike a jarring note if I add that earlier in his *History* Sprat had said: "The first thing

5. Thomas Sprat, *The History of the Royal Society of London for the Improving of Natural Knowledge*, 1667, p. 392.
6. *Ibid.*, p. 393.

that ought to be improved in the English nation is their industry. . . . A true method of increasing industry is by that course which the Royal Society has begun in philosophy, by works and endeavours, and not by the prescriptions of words or paper commands."[7]

Sprat's views are very understandable in their context because mechanical industry was growing apace in England: we had been having our first Industrial Revolution. What is more surprising, perhaps, is that Samuel Taylor Coleridge, in his Introduction to the *Encyclopaedia Metropolitana,* wrote the following expostulation: "It is not, surely, in the country of ARKWRIGHT, that the Philosophy of Commerce can be thought independent of Mechanics: and where DAVY has delivered Lectures on Agriculture, it would be folly to say that the most Philosophic views of Chemistry were not conducive to the making our valleys laugh with corn."

The most sinister consequence of looking down on applied science was a backlash that has diminished pure science in favor of its practical applications and that culminated in England in the injudicious advocacy that sought to fund research on the basis of the retail trade: the so-called consumer-contractor principle. The pejorative use of the word *academic*—found only among the lowest forms of intellectual life—became quite common. Sprat would have thought such a turn of opinion very strange, as he said in writing of the Royal Society:

It is strange that we are not able to inculcate into the minds of many men, the necessity of that distinction of my Lord Bacon's, that there ought to be Experiments of Light, as well as of Fruit. It is their usual word, *What solid good will come from thence?* They are indeed to be commended for being so severe Exactors of goodness. And it were to be wished, that they would not only exercise this vigour, about Experiments, but on their own lives and actions: they would still question with themselves, in all that they do; *What solid good will come from thence?* But they are to know, that in so large and so various an Art as this of Experiments, there are many degrees of usefulness: some may serve for real, and plain benefit, without much delight: some for teaching without apparent profit: some for light now, and for use hereafter; some only for ornament, and curiosity. If they will persist in contemn-

7. *Ibid.,* p. 421.

ing all Experiments, except those which bring with them immediate gain, and a present harvest: they may as well cavil at the Providence of God, that he has not made all the seasons of the year, to be times of mowing, reaping and vintage.[8]

It *is* strange, is it not?

The Critical Mind

A scientist who wishes to keep his friends and not add to the number of his enemies must not be forever scoffing and criticizing and so earn a reputation for habitual disbelief; but he owes it to his profession not to acquiesce in or appear to condone folly, superstition or demonstrably unsound belief. The recognition and castigation of folly will not win him friends, but it may gain him some respect.

Over a period of years I have collected a little treasury of more or less fallacious beliefs, and a discussion of some of these will help to exemplify criticisms of the kind I think just.

How often has it not been contemptuously said that "modern medicine cannot even cure the common cold"? What is offensive here is not the statement's falsity (it is true) but its implication: is it not pointless to pour billions of dollars into cancer research when modern medicine . . . and so on. What is wrong here is the almost universally held belief that clinically mild diseases have simple causes while grave diseases are deeply complex and are proportionately difficult to discern the causes of or to cure. There is no truth in either; a common cold, caused by one or more of a multiplicity of upper respiratory viruses and with an overlay of allergic reactivity, is an extremely complex ailment; so is eczema, most forms of which are baffling still. On the other hand, some very grave diseases such as phenylketonuria have relatively simple origins; some can be prevented, as phenylketonuria can be, or cured, as so many bacterial infections can be. Some forms of cancer are simple in origin and can be circumvented—for example, the cancers caused by smoking and by certain industrial chemicals. Indeed, good judges have put the proportion of all can-

8. *Ibid.*, p. 245.

cers caused by extrinsic agents as high as 80 percent.

Another declaration of the same genre as the one about the common cold is that "cancer is a disease of civilization"—a seemingly natural inference from the fact that cancer is much more widely prevalent in the industrialized countries of the Western world than in the developing nations. But it is second nature to people familiar with demography or epidemiology to ask if the populations being compared are genuinely comparable, and here they are not. It is his relatively high expectation of life—of his not dying, that is to say, for other reasons—that confers upon Western man his relatively high expectation of contracting cancer—a disease of middle or later life—so the inference is unsound. Comparisons of mortality are valid only if populations are standardized with respect to variables such as their age compositions, with allowance, too, for differences in skills of diagnosis.

Another way in which a scientist can lose friends is to call attention to the tricks that selective memory can play upon judgment. "Three times, no less, I dreamed of Cousin Winifred, and on the very next day she rang me up. If that doesn't prove that dreams can foretell the future, then I'm sure I don't know what does." But, the young scientist expostulates, on how many occasions did you dream of Cousin Winifred without a subsequent telephone call? And is it not a fact that she rings up almost every day? We remember only the striking conjunctions; there is no incentive to remember occasions when misfortunes come singly or in pairs, and not in threes (or whatever other number superstition fixes on). Seeing an example of bad driving, a man of a certain temperament will remark it and remember it only if the car is driven by a woman—and thus he convinces himself of women's lesser skill without realizing his own errors of judgment.

Writing on the same subject as this, the endocrinologist Dr. Dwight Ingle has recounted the following variant of a chestnut of immemorial origin:

PSYCHIATRIST: Why do you flail your arms around like that?
PATIENT: To keep the wild elephants at bay.
PSYCHIATRIST: But there aren't any wild elephants here.
PATIENT: That's right. Effective, isn't it?

Post hoc, ergo propter hoc has many devotees, and some of them, I fear, have been scientists. Classical embryologists, for example, were at one time wont to believe that a complete anatomical record of antecedent states provided causes enough to explain development.

Superstitions are not so easy to cope with. Probably it is better not to try to reason with astrological predictions, but it may be worthwhile just once to call attention to the extreme *a priori* unlikelihood of their being true, and point to the lack of any convincing evidence that they are so. But perhaps after all it is best to let sleeping unicorns lie—I myself have for some time past abstained from discussing spoon bending or other manifestations of "psychokinesis."

Wise scientists and medical men take some pains to guard against the dangers arising out of a predilection for getting one experimental result rather than another. If an experiment cannot be fully controlled, matters are so arranged that uncontrollable sources of error tell, if anything, *against* the hypothesis one would like to see corroborated. Moreover even the most experienced and honorable clinicians fall in readily with the drill of "double-blind trials"—those in which neither physician nor patient knows whether the patient has received a supposedly efficacious medicament or a placebo made up to look and taste like it. If strictly carried out, and if the member of the team responsible for it has not lost the key of the code, evaluation of the treatment can be carried out on a genuinely objective basis, uninfluenced by the physician's wishes, or the patient's.

Exaggerated claims for the efficacy of a medicament are very seldom the consequence of any intention to deceive; they are usually the outcome of a kindly conspiracy in which everybody has the very best intentions. The patient wants to get well, his physician wants to have made him better, and the pharmaceutical company would like to have put it into the physician's power to have made him so. The controlled clinical trial is an attempt to avoid being taken in by this conspiracy of good will.

7

Of Younger and Older Scientists

Youth, though endearing, has its pitfalls, and no work of this nature could be complete unless it drew special attention to them.

Excess of Hubris. Success has sometimes a bad effect on young scientists. Quite suddenly it turns out that everyone else's work is slovenly in design or incompetently carried out; the young genius won't accept it until he has "looked into it himself." Yes, certainly he will give a paper at the next meeting of the society. True, he gave a paper at the last meeting, but things have moved on since then, and a whole lot of people will be anxious to hear about these later developments.

The old-fashioned remedy for hubris was a smart blow on the head with an inflated pig's bladder—and this is in the spirit of the rebuke that may have to be administered before the young scientist injures himself in the opinions of those who would otherwise like him and wish him well.

Brilliant Young Scientist. While he *is* young and if he is genuinely brilliant, his colleagues will exercise forbearance and may even feel affectionate pride at the manifestations of the razor-sharp intellect, the lightning comprehension, and the uncanny facility with which he recollects facts or notions recorded only in the *Proceedings of the National Academy of Sciences* of a banana republic or in a long-out-of-date issue of *The Grocer and Fishmonger.*

Ambition. Considered as a motive force that helps to get things done, ambition is not necessarily a deadly sin, but excess of ambition can certainly be a disfigurement. An ambitious young scientist is marked out by having no time for anybody or anything that does not promote or bear upon his work. Seminars or lectures that do not qualify are shunned, and those who wish to discuss them are dismissed as bores. The ambitious make too obvious a point of being polite to those who can promote their interests and are proportionately uncivil to those who cannot. "I hope we don't have to be nice to *him,*" an ambitious young Oxford don said to me of a kindly old buffer with an amateurish interest in science who was dining at High Table. He wasn't, and although this particular episode did not harm him, it was symptomatic of a state of mind that did.

Growing Older

Like any other human being, a young scientist growing up will probably say to himself at the end of each decade, "Ah well, that's it, then. It has been great fun, but nothing now remains except to play out time with dignity and composure and hope that some of my work will last a bit longer than I do."

Such dark thoughts are wider of the mark with scientists than with most other people. No working scientist ever thinks of himself as old, and so long as health, rules of retirement, and fortune allow him to continue with research, he enjoys the young scientist's privilege of feeling himself born anew every morning. This infectious zest was one of the most endearing characteristics of that great generation of American biologists on whose behalf all ordinary laws of mortality and even the physical intimations of it seemed to have been held in abeyance: Peyton Rous (1879–1970), G. H. Parker (1864–1955), Ross G. Harrison (1870–1959), E. G. Conklin (1863–1952), and Charles B. Huggins (1901–).

The whole question of which faculties deteriorate most rapidly with increasing age has not yet been sufficiently explored. It is an easy assumption that creativity deteriorates sharply. The octogenarian Verdi of *Falstaff* is usually called to the witness stand to rebut it, and when he steps down, the Titian of the later great paintings testifies with equal conviction. It is not true that "research is a young man's game" nor that high awards are won with disproportionate frequency by the young. Harriet Zuckerman has shown in *Scientific Elite,* her study of American Nobel laureates, that in relation to the population "at risk," as actuaries say—at risk of making a contribution to science—the modal age at which laureates did the work that won them their prizes was early middle age.

I am sorry to say that when I think of older scientists the picture that forms in my mind is of a committee of grayheads, all confident in the rightness of their opinions and all making pronouncements about the future development of scientific ideas of a kind known by philosophers to be intrinsically unsound.[1]

In my middle age I became great friends with Sir Howard Florey, my first boss, who had earlier developed the fungal extractive penicillin. Florey very greatly resented the time and energy he had to spend finding funds to support his research. He had applied for help to a committee of high-ups who might have been expected to give it; but no, wise old gray heads shook ("or perhaps merely wobbled," Florey said) from side to side as they pronounced that the future of antibacterial therapy lay with synthetic organic chemicals of which Gerhard Domagk's sulfanilamide was the paradigm, and certainly not with fungal or bacterial extractives that seemed to belong to the pharmacopoeia of *Macbeth,* act 4, scene 1. An historian of the body of high-ups to which I am referring defended them to me privately by saying that the view they took was a perfectly reasonable one at the time, but this is not an adequate defense. Nor should it have been relevant—though in real life these things are—that Florey was an impatient and aggressively confident man; the committee were at fault because they formed a confident judgment in a context in which nothing but the most hesitant and tentative opinion was justifiable.

What I find unforgivable is that the opinion about sulfona-

1. See P. B. Medawar, "A Biological Retrospect," in *The Art of the Soluble* (New York: Barnes and Noble, 1967), especially p. 99, which starts with a formal refutation of the notion that future ideas can be predicted.

mides and synthetic organic chemicals generally was so utterly unimaginative and lacking in shrewdness. Although the Official Secrets Act draws an impenetrable veil over the whole proceeding, I can so easily imagine the knowing way in which members of this committee assured each other of the truth of the utterly banal view that one day (ah, but when—the war was on, after all) synthetic organic chemicals would sweep aside the brews that biologists take such pains to prepare. For all I know, though, the genuinely wiser members of the committee may have thought that Florey's and Fleming's ideas were worth trying, but were talked down by someone with so confident an air and so assertative a voice they were ashamed of being classified as fuddy-duddies, the champions of an ostensibly old-fashioned view.

Excess of confidence in the rightness of their own views is a sort of senile hubris, as offensive in older scientists as excess of hubris in the young.

The tenor of the above remarks will at once be recognized as grossly unfair by those who reflect that money for research is limited and a choice between one enterprise and another must be made. That is all very well; but it is not so much the wrongness of the judgments as their pretensions to rightness that attracts the animosity of younger scientists, just as the professional tipsters or soothsayers are blamed not so much for the falsity of their prognostications as for the claim that they will be right. A senior scientist in a position of great responsibility should always hear behind him a voice such as that which reminded a triumphant Roman emperor of his mortality, a voice that should now remind a scientist how easily he may be, and how often he probably is, mistaken. When I had spent a few years in his laboratory, Professor Florey complained to me that he then seemed to spend most of his time making it possible for others to undertake research, but it was characteristic of his real kindness and gruff common sense that he saw it as a principal function of the older scientist to promote the welfare of the young.

When dealing with older scientists, the young should not assume that their elders remember their names or still less their faces, notwithstanding that friendly chat on the boardwalk of

Atlantic City at the federation meetings[2] as recently as a year beforehand.

Nor should the young attempt to ingratiate themselves with their seniors; such attempts miscarry so often that they had better be abjured.

A senior scientist is much more flattered by finding that his views are the subject of serious criticism than by sycophantic and sometimes obviously simulated respect. A young scientist will not, however, ingratiate himself with a prospective patron by exposing his views to scathing public criticism. Older scientists expect nothing more from the young than civility. Cobbett was very firm about the evils of "sucking up": "Look not for success to favour, partiality, and friendship or to what is called *interest:* write it upon your heart, that you will depend solely on your merit and your own exertions."

For their part, older scientists must remember—what I constantly forget—that not even the most brilliant of their juniors can remember the great stir caused by O. T. Avery's announcement that the type transformation of pneumococci was mediated through the action of DNA. Most of today's graduate students weren't born in 1944, anyway, and events that happened as long ago as that are thought by the young to belong to a pre-Cambrian era of scientific growth. The young, moreover, can tire of hearing what a remarkable fellow old Dale was, what a card Astbury, and how cruelly adept J. J. Thompson at putting his juniors down. The young scientist will find, though —as Lord Chesterfield could have told him—that if he simulates interest in these yarns, he may become interested in spite of himself and learn something that may improve his mind.

Even if the motive is one of self-congratulation, we all think it natural and agreeable when an old buffer says, "I was immensely bucked when I saw that Wotherspoon had won this year's chemistry prize: he was my pupil, you know"—you do now, anyway—"and even in those days he was as bright as a

2. The annual convention of the American Societies of Experimental Biology, often held in Atlantic City, is an enormous concourse attended by thousands of scientists, at which senior scientists may seek likely recruits and youngsters may hope to attract patrons.

new penny." This generous attitude of mind is not universal because for complex psychological reasons some tutors and supervisors are well known habitually to eat their young.

Looking at the same relationship from the other end of life, I believe an attitude of friendly respect toward their mentors is proper in the young. No remark is more disagreeable from young lips than "I am sorry old Wotherspoon has died, of course, but he never really was any good you know"—you did not. Lord Chesterfield would have been inexpressibly shocked by such a declaration. If thought, such sentiments should be left unspoken.

Science and Administration

Young scientists wishing to be thought even younger and more inexperienced than they really are should lose no opportunity to jibe at and belittle the administration, whatever it may be. It would help them to grow up if they realized that scientific administrators are problem-solvers as they are—and are working, too, for the advancement of learning. In some ways, a young scientist should reflect, the administrator's task is the more difficult, for whereas well-established laws of nature discourage a young scientist from attempting to circumvent the Second Law of Thermodynamics, no comparable body of administrative common law assures the administrator that he can't get a quart into or out of a pint pot, or money out of a stone —feats executed or attempted daily by administrators trying to raise funds. Nor can they turn barren ground overnight into sumptuously equipped laboratories.

Young scientists may be wrong to assume that scientific administrators who have had scientific careers themselves will necessarily be the most sympathetically attentive to their needs; for having been scientists and therefore supplicants themselves, they are likely to know all the tricks for trying to raise funds—and in particular the argument that if only the work currently in progress could be prolonged for a few years, it would dizzily expedite our understanding of the etiology of cancer or of the mechanism of cell division.

A senior scientist usually turns to administration because he believes that this is the best way he can contribute to the advancement of learning—which is, or ought to be, a young scientist's ambition, too. Such a decision cannot be made without personal sacrifice; very often it means giving up research, for major administrative jobs are too demanding to make it possible to continue an activity that calls for the almost obsessional single-mindedness required by almost any human endeavor that is to be well and quickly done, including administration itself.

Young scientists must on no account complain that they don't have enough say in things and then complain even more when they are invited to serve on committees that will give them the say they think they ought to have. Service on committees, young scientists will find, eats up time they would really much prefer to spend in the laboratory in spite of all those complaints about administrators who push them around. Because of the growing importance of science, scientific administration is now a job as important and as well defined as the administration of hospitals, and no physician or surgeon would think to lay aside his stethoscope or scalpel to do the work of the almoners or engineers—he lets the administration get on with it. A young scientist should do the same; if he rates administration so low, he should think himself lucky not to be engaged in it.

Service on committees and other extramural distractions should never be used as an excuse for not doing research, for that is the scientist's first business. I know no good scientist who makes such excuses—only bad ones. So great is the counterappeal of laboratory work that the burden of administrative duties upon a scientist is almost always overestimated. I knew an able young colleague who left a famous university to take a commercial job in a pharmaceutical laboratory. I asked him how he liked the change, and he declared himself delighted—university administration had been "getting a bit on top of him." Unaware that he had had any administrative duties at all, I asked him what his administrative work had been. "Oh well," he said with an air of graduate martyrdom, "I was roped onto the wine committee, you know." It had been an excellent appointment, too.

The conciliatory and magnanimous tone of my remarks on

administration could be construed as the testimony of a reformed drunkard now bounding after his former drinking companions with the pledge. For his part, no scientific administrator should lose sight of Haddow's[3] Law: it is the administrator's job to get money and the scientist's to spend it.

Although it is thought to be true (I quote Stella Gibbons's mock-Lawrence from *Cold Comfort Farm*) that there must always be a deep dark bitter belly-tension between scientists and administrators, one of the benefactions of increasing age and experience is the realization that everyone gets on better if a generally matey atmosphere prevails.

Time Needed for Reflection. I can remember my seniors saying as they hurried off with an air of martyrdom to attend meetings of committees to which they need never have belonged, "I never seem to get any time for thinking nowadays." I found this remark puzzling because it did not seem to me to be possible to apportion a time for thinking, as for playing squash, dining, or having a drink.

What they meant was that they had no time for reading of cognate but not directly relevant scientific literature, for reflection, for the unhurried musing over experimental results—their own and others'—looking for unsuspected sources of error and wondering upon the new directions the research might take. A scientist who is deeply preoccupied with the solution of a problem will find not so much that he allocates special times to thinking about it but rather that reflection upon the problem is the equilibrium state or the zero point on the dial to which his mind tends automatically to return when it is not occupied by anything else. Indeed, when a scientist without administrative responsibilities is very deeply engaged in his research, the problem is not so much to find time for reflection on his research as to find time for not reflecting upon it and doing instead any one of the hundred other things that good parents, spouses, householders or citizens should be concerned with.

---

3. Sir Alexander Haddow, for many years the head of Britain's largest cancer research institute, the Chester Beatty.

8

Presentations

Scientific research is not complete until its results have been made known. With scientists, publication almost always takes the form of a "paper" written for a learned journal—in contrast to humanists, who often publish their research in the form of books. It is because scientists so seldom write books that old-fashioned humanists—such as are still to be found in Oxford or Cambridge colleges—are sometimes inclined to question their productivity and wonder whether the long hours in a laboratory are not devoted to hobbies or to some form of play.

The delivery of a paper to a learned society is a form of publication but is not thought definitive until it appears in print. At some stage in his life, a young scientist will inevitably have to give a paper to a learned society, though not before he has tried it out on his mates at, for example, a departmental seminar. This is a friendly and relaxed occasion, but a paper to a learned society requires a little more address. *Under no circumstances whatsoever should a paper be read from a script.* It is hard to overestimate the dismay and resentment of an audience that has to put up with a paper read hurriedly in an even monotone. Speak from notes, young scientist; to speak without is a form of showing off and only creates the impression (perhaps well-founded) that the same story has been told over and over again. Notes should be brief and never consist of long paragraphs of stately prose. If a few cues aren't sufficient to get a speaker into motion, then he must go over the topic repeatedly—not necessarily aloud—until the right words come at the appropriate stimulus. I early found it to be a great help when trying to expound a difficult concept to write "(EXPLAIN THIS)" after it appeared in the notes—a device that of course forces a speaker to find natural words.

A torrential outpouring of words may make the speaker think that he is being very brilliant, but his audience are more likely to think him glib. A measured delivery with perhaps a touch of gravity is what Polonius would surely have recommended. Try also not to be a bore. A scientist who takes time off to lecture to schoolchildren will soon learn whether or not he has his audience in the palm. Children cannot keep still, and if they are bored, they fidget. The lecturer may sometimes feel he is addressing an enormous audience of mice, but the moment the very young are interested, they sit still.

A lecturer can be a bore not only by being insufferably prosy or because his work is intrinsically dull, but because he goes into quite unnecessary details about matters of technique. Sometimes it is judicious to spare an audience the details. If it is important to know and if the audience wants to know the order in which the speaker dissolved the various ingredients of his nutritive culture medium, he will be asked immediately after the lecture or privately later on.

Whenever possible, the blackboard should be used in preference to slides; I have presided over very successful conferences in which all lantern slides and formal orations were prohibited. Such considerations do not apply, of course, when the *exact* forms of a curve or of a family of curves are crucially important, or the exact numerical values of a set of radioactive counts. Very often they are not; if the relationship between the variables is linear—one of simple proportionality—then say so. If the audience won't take a scientist's word for it, they won't take his slide for it, either. If the speaker's contention is challenged, it will only be necessary for him smugly to ask the projectionist, "Would you please be kind enough to show slide seven?"—that which will show beyond a peradventure that the relationship is linear indeed.

Length is a problem. Speakers should remember a principle of almost Newtonian stature, which I believe was first propounded independently and on the same occasion by Dr. Robert Good and myself: that people with anything to say can usually say it briefly; only a speaker with nothing to say goes on and on as if he were laying down a smoke screen.

Of all the monsters of science fiction the Boron is that which arouses the greatest dread—anyhow at scientific conferences. There is, incidentally, no more expeditious way of making a lifelong enemy than to poach upon the next speaker's ration of time—something that should never happen anyway if the chairman is awake.

Even the most experienced speakers feel nervous before a talk, and it is very right that they should do so, for it is a sign of anxiety to do well. Audiences are not really impressed when a lecturer rummages in his pocket for a crumpled envelope and says (as I once heard J. B. S. Haldane say), "When I was wondering on the train what I should say to you . . ." Audiences respond better to evidence that the speaker has been at pains to prepare whatever he has to say. Lantern slides illustrating the lecturer's fingerprints or the fracture patterns of glass must be shunned.

The most difficult rule of self-discipline is to learn not to get flustered if misadventures occur—as sometimes, inevitably, they will. An audience is more indulgent to a speaker who loses his place, muddles his slides, or even falls off the rostrum than to one who has given any evidence of treating them with less than due respect.

Not very long after a severe illness that impaired my eyesight and cost me the use of one hand, I got sadly muddled with my speaking notes at a big public lecture. My wife came to the platform to help, and the audience, who had been suffering vicariously, as nice people do, were delighted and relieved to hear me say to her over the public-address system, "I see exactly what you mean—page five comes after page four."

In Great Britain, the Institution of Electrical Engineers issues an admirable *Speaker's Handbook* in which a speaker is recommended to stand with his feet 400 millimeters apart, as "this stops trembling." The instruction is amusing not because electrical engineers are especially tremulous but because of its high degree of quantitative refinement—it is as if experiments had shown that feet 350 or 450 millimeters apart could be relied upon to precipitate a bout of convulsions.

Scientists should behave in lectures as they would like others to behave in theirs. It is an inductive law of nature that lecturers always see yawns and *a fortiori* those hugely cavernous yawns that presage the almost complete extinction of the psyche. The same goes for anything else that may distract a lecturer (which may, of course, be the intention): sibilant whisperings, ostentatious consultations of watches, laughter in the wrong places, slow, grave shakings of the head, and the like. A member of the audience thought to be an expert on the topic of the speaker's discourse is well advised to think of a question to ask in case the chairman turns to him and says, "Dr. ——, we have just a few moments for discussion, so why don't you set the ball rolling?" The person to whom this invitation is addressed cannot very well say, "I'm afraid I can't—I was fast asleep," but if he merely says, "What do you envisage as the next step in your research?" the audience will take it for granted that he was. Sleepiness is quite often due to hypoxia in a badly ventilated lecture room —*not* necessarily to boredom.

If people *do* sleep in their lectures, speakers should try to get some comfort from the thought that no sleep is so deeply refreshing as that which, during lectures, Morpheus invites us so insistently to enjoy. From the standpoint of physiology, it is amazing how quickly the ravages of a short night or a long operating session can be repaired by nodding off for a few seconds at a time.

Writing a Paper

No number of lectures, seminars or other verbal communications can take the place of a contribution to a learned journal. It is well known, though, that the prospect of writing a paper fills scientists with dismay and brings on a flurry of displacement activities: uselessly uninformative experiments, the building of functionless or unnecessary apparatus, or even, *in extremis,* attendance at committees ("If I don't occasionally attend the security committee, everyone will think that I'm the thief"). The traditional reason given for a scientist's reluctance to write a paper is that it takes time away from research; but the real explanation is that writing a paper—writing anything, indeed, even the begging letters that are necessary if a laboratory is to remain solvent—is something most scientists know they are bad at: it is a skill they have not acquired.

Scientists are supposed to have an intuitive ability to write papers because they have consulted so many, just as young teachers are supposed to be able to give lectures because they have so often listened to them.

I feel disloyal but dauntlessly truthful in saying that most scientists do *not* know how to write, for insofar as style does betray *l'homme même,* they write as if they hated writing and wanted above all else to have done with it. The only way to learn how to write is above all else to read, to study good models, and to practice. I do not mean to practice in the sense in which young pianists practice "The Merry Peasant," but practice by writing whenever writing is called for, instead of making excuses for not doing so, and writing, if necessary, over and over again, until clarity has been achieved and the style, if not graceful, is at least not raw and angular. A good writer never makes one feel as if one were wading through mud or picking one's way with bare feet through broken glass. Further, writing should be as far as possible natural—that is, not worn like a Sunday suit and not too far removed from ordinary speech, but rather as if one were addressing one's departmental chairman or other high-up who was asking about one's progress.

No number of "don'ts" will make a "do," but certain practices should certainly be shunned. One such was introduced into American English from Germany—that of using nouns attributively (as if they were adjectives), sometimes stringing them all together to make one huge nounlike monster in constant danger of falling apart. A skillful linguist but habitual liar once told me of a single word in German standing for "the widow of the man who issued tickets at reduced prices for admission on Sundays to the zoo." This is untrue, of course, but it illustrates the principle, and if I myself have not read about "vegetable oil polyunsaturated fatty acid guinea pig skin delayed type hypersensitivity reaction properties," I have read some equally daunting nounal phrases. An incentive to write like this is that most editors restrict the length of a paper, so that a scientist who makes one word do the work of ten may feel he is one up on the editor.

Another little rule (for medical scientists especially) is that mice, rats, and other laboratory animals should never be injected. Few hypodermic needles are large enough for even the smallest mouse to pass through, especially if it is injected with something. ("Mice were injected with rabbit serum albumin mixed with Freund's adjuvant," we read. "Ah, but what into?" the cry goes up.) Mice should receive injections, or substances should be injected into them. Preciosity? Considered in isolation, yes, but it is the accumulation of such errors of taste that disfigures what could otherwise be a straightforward and readable paper. Avoid, too, such weary tropes as "the role of (or the part played by) adrenocortical hormones in immunity." Why not write instead "the contribution of adrenal cortical hormones to . . ." and so on. Give thought to prepositions, too: the regulation of electrolytes in the body is mediated not *by* but *through* the adrenal gland. Again, we are (or are not) tolerant *of,* not tolerant *to,* errors of literary judgment, and so on.

Another thought to bear in mind is that good writing upon a subject is almost always shorter than bad writing on the same subject. It is often much more memorable, too. Who but Winston Churchill could have said so much in so few words as my Lord Bacon's comment on an ambitious political rival: "He doth like the ape, that the higher he clymbes the more he shows his ars [*sic*]."

But if a young scientist is to study models, which are they to be? Any technically skillful writer will do, especially if it is a writer the reader admires and would like to read anyway. Fiction and other nonexpository writing will do very well; Bernard Shaw wrote a very good sentence, and some of Congreve's writing is miraculously skillful, but I especially recommend the writing of those who are expounding difficult subjects and are determined to make themselves understood. Although not all philosophers satisfy this requirement, they are in the main an excellent choice, particularly, I believe, those who have been professors of philosophy in University College London; A. J. Ayer, Stuart Hampshire, Bernard Williams, and Richard Wollheim are among them. Essayists are often good models; Bacon's essays are superlative, and some of Bertrand Russell's essays (for example, his *Sceptical Essays*) are brilliantly well written. So are many of J. B. S. Haldane's, now mostly out of print. Gravity, wit and a strong understanding have never been more effectively combined than in Dr. Johnson's *Lives of the Poets.*

In the English-speaking world (people think differently about these things in France), scientific and philosophic writing is never now allowed to be an exercise in the high rhetoric style, but in the days when there was still an element of conflict between style and substance or medium and message, Dr. Joseph Glanvill, F.R.S. (1636–80) thought it right to put natural philosophers on their guard. A scientist's writing, he wrote in *Plus Ultra,* was to be "manly and yet plain . . . polite [polished] and as fast as marble." It was not to be "broken by ends of Latin nor impertinent quotations, . . . not rendered intricate . . . by wide fetches and circumferences of speech."

Most of these cautions are no longer relevant, nor is Abraham Cowley's advice in his ode to the Royal Society to abjure "the painted scenes and pageants of the brain." Long wavy hair was out then, and the close crop appropriate to the radical Puritan activists who played such a large part in auguring the scientific revolution was to be the fashion of the day. Consider, for example, the opening paragraph of Bertrand Russell's *Sceptical Essays*—that in which he outlines his intentions. It is hard to imagine writing clearer, more pointed, or more succinct; notice also how very like it is to speech—one can nearly hear again that dry crackly Voltairean voice:

I wish to propose for the reader's favourable consideration a doctrine which may, I fear, appear wildly paradoxical and subversive. The doctrine in question is this: that it is undesirable to believe a proposition when there is no ground whatever for supposing it true. I must, of course, admit that if such an opinion became common it would completely transform our social life and our political system; since both are at present faultless, this must weigh against it. I am also aware (what is more serious) that it would tend to diminish the incomes of clairvoyants, bookmakers, bishops and others who live on the irrational hopes of those who have done nothing to deserve good fortune here or hereafter. In spite of these grave arguments, I maintain that a case can be made out for my paradox, and I shall try to set it forth.

In writing a paper, a young scientist should make up his mind about whom he is addressing. The easy way out is to address one's professional colleagues only—and of them, only those who work in a field cognate with one's own. This is not at all the way to go about it. A scientist should reflect that his more intelligent peers probably browse in the literature for intellectual recreation and might like to find out what he is up to. The time will come, moreover, when a young scientist will be judged upon his written work by referees and adjudicators. They are entitled to feel annoyed—and often do—when they can't make out what the paper is about or why the author undertook the investigation, anyway. A formal paper should therefore begin with a paragraph of explanation that describes the problem under investigation and the main lines of the way the author feels he has been able to contribute to its solution.

Great pains should be taken over the paper's summary, which should make use of the whole of the journal's ration of space (one-fifth or one-sixth of the length of the text, as the case may be), and its composition is the severest test of an author's literary skill, particularly in days when "précis writing" has been dropped from the syllabus in most schools for fear of stifling the scholars' creative afflatus. The writing of a summary tests the author's powers of apprehension and sense of proportion—the feeling for what is really important and what can be left out. A summary must be complete in its own limits. It may well start with a statement of the hypothesis under investigation and end with its evaluation. Nothing is more abjectly feeble than to write some such sentence as "The relevance of these findings to the etiology of Bright's disease is discussed." If it *has* been discussed, the discussion should be summarized, too. If not, say nothing. The preparation of abstracts is a public service a young scientist should sometimes volunteer to do. Even if his work is overseen by an experienced editor before it goes to press, abstracting can be good practice in writing.

The number of references cited in the literature list (be always scrupulously careful to observe the house style) should be that which is sufficient and necessary; it may be a symptom of scientmanship (see Chapter 6) to quote references from journals published so long ago that librarians desperate for space have long since had them stashed away in the galleries of disused mines. Due homage and justice to one's predecessors are criteria to keep in mind, although some names are so great and some ideas so familiar that omission is homage greater than citation. Nice judgment is needed, though; one man's compliment may be another's source of grievance.

Papers embodying good work may be rejected by an editor for a variety of reasons. Publishers of scientific journals like it to be known that they are being beggared by the prolixity of their contributors, and a length disproportionate to content is indeed the most common cause of rejection. Another is citation in the literature list of papers not referred to in the text or vice versa. In such a case, rejection is condign. Whatever the reason given for it, rejection of a paper is always damaging to the pride, but it is usually better to try to find another home for it than to wrangle with referees. There are times when referees are inimical for personal reasons and enjoy causing the discomfiture that rejection brings with it; too strenuous an attempt to convince an editor that this is so may, however, convince him only that the author has paranoid tendencies.

Of the internal structure of a paper I have said only that one should have a first explanatory paragraph describing in effect the problem that is preying on the author's mind. The layout of the text that has come to be regarded as conventional is that which perpetuates the illusion that scientific research is conducted by the inductive process (see Chapter 11). In this conventional style, a section called "Methods" describes in sometimes needless detail the technical procedures and reagents the author has used in his research. Sometimes a separate section headed "Previous Work" may concede that others have dimly groped their way toward the truths the author is now proposing to expound. Worst of all, a paper in the conventional layout may contain a section called "Results"—a voluble pouring forth of factual information, usually with no connecting narrative to explain why one observation is made or one experiment done rather than another. Then follows a passage called "Discussion" in which the author plays out the little charade that he is now going to collect and sort out all the information he has gathered by wholly objective observation with the purpose of finding out what, if anything, it means. This is the reductio ad absurdum of inductivism—a faithful embodiment of the belief that scientific inquiry is a compilation of facts by the contemplation or logical manipulation of which an enlargement of the understanding must inevitably follow. This division of "Results" from "Discussion" may be thought to have its parallel in the praiseworthy editorial policy of those reputable newspapers which divide news from editorial comment upon it, but the two cases are in no way parallel; the reasoning that is called "Discussion" in a scientific paper is in real life integral with the process of securing information and having the incentive to do so. The separation of "Results" from "Discussion" is a quite arbitrary subdivision of what is in effect a single process of thought. Nothing of the kind applies to the dissociation of news of events or legislative action from editorial comment upon them, for these two can vary independently.

A scientist who completes writing—or, as people unaccountably say, "writing up" a paper (by which, of course, they mean "writing down")—should feel proud of it, should feel, indeed, "this will make people sit up." It shows either a poor spirit or perhaps good judgment if no such thought enters the author's head.

When I was director of the National Institute for Medical Research, a young colleague of mine completed a brief letter to *Nature*—the traditional vehicle of important scientific news—that was so important, he felt, and so eagerly awaited by the world that it should not be entrusted to the post but must be delivered by hand. So it was. But then, unfortunately, it was lost and had to be resubmitted. This time, it went by post. We all felt that on the previous occasion it had been pushed under the door and therefore probably ended up under the welcome mat. *Moral:* use the recognized channels of communication.

9

Experiment and Discovery

Ever since Bacon's day experimentation has been thought to be so deeply and so very necessarily a part of science that exploratory activities that are not experimental are often denied the right to be classified as sciences at all.

Experiments are of four kinds;[1] in the original Baconian sense, an experiment is a contrived, as opposed to a natural, experience or happening—is the consequence of "trying things out" or even of merely messing about.

The reason why Bacon attached such great importance to experiments of this kind is explained later, but it was of Baconian experiments—those that answer the question "I wonder what would happen if . . ."—that Hilaire Belloc must have been thinking when he wrote the following passage:

> Anyone of common mental and physical health can practise scientific research. . . . Anyone can try by patient experiment what happens if this or that substance be mixed in this or that proportion with some other under this or that condition. Anyone can vary the experiment in any numbers of ways. He that hits in this fashion on something novel and of use will have fame. . . . The fame will be the product of luck and industry. It will not be the product of special talent.[2]

---

1. In this chapter I shall be following and explaining more fully than hitherto the scheme of classification proposed in my *Induction and Intuition in Scientific Thought* (Philadelphia: American Philosophical Society, 1969).

2. Quoted from an admirable anthology of quotations to do with science: Alan L. Mackay, *The Harvest of a Quiet Eye* (Bristol: Institute of Physics, 1977).

Baconian Experimentation. In the early days of science,[3] it was believed that the truth lay all around us—was there for the taking—waiting, like a crop of corn, only to be harvested and gathered in. The truth would make itself known to us if only we would *observe* nature with that wide-eyed and innocent perceptiveness that mankind is thought to have possessed in those Arcadian days before the Fall—before our senses became dulled by prejudice and sin. Thus the truth is there for the taking if only we can part the veil of prejudice and preconception and *observe things as they really are;* but alas, we might spend a whole lifetime observing nature without ever witnessing those conjunctions of events that could reveal so much of the truth if by chance they came our way. It is no use, Bacon explained, relying upon good fortune—on "the casual felicity of particular events"—to furnish us with all the factual information we need for apprehending the truth, so we must *devise* happenings and contrive experiences. In John Dee's words, the natural philosopher must become the *"archmaster"* who *stretches* experience. The "electrification" of amber by rubbing and the communication of magnetic properties to iron nails from a lodestone are good examples of the experiments Bacon advocated; again, we know what happens if we distill fermented liquors once, but what happens if we distill the distillate a second time? Only by experimenting in this fashion can we build up that majestic pile of factual information from which, according to the mistaken canon of inductivism (see Chapter 11, "The Scientific Process"), our understanding of the natural world will necessarily grow.

It may have been their perseverence in experimentation of this kind—often involving messy manipulations and even offensive smells—that caused scientists to be looked down upon by the genteel.

Aristotelian Experiments. In explaining this second kind of experimentation I have followed a lead of Joseph Glanvill's. This experiment, too, was contrived—to demonstrate the truth

3. K. R. Popper, "On the Sources of Knowledge and of Ignorance," in *Conjectures and Refutations* (New York: Basic Books, 1972).

of a preconceived idea or to act out some calculated pedagogic plot: apply electrodes to the frog's sciatic nerve, and lo, the leg kicks; always precede the presentation of the dog's dinner with the ringing of a bell, and lo, the bell alone will soon make the dog dribble. Joseph Glanvill, in common with many of his contemporary Fellows of the Royal Society had the utmost contempt for Aristotle, whose teachings he regarded as major impediments to the advancement of learning. In *Plus Ultra* he wrote of such experiments thus: "Aristotle . . . did not use and imploy Experiments for the erecting of his Theories: but having arbitrarily pitch'd his Theories, his manner was to force Experience to suffragate, and yield countenance to his precarious Propositions."

Galilean Experiments. Neither the Baconian nor the Aristotelian but rather the Galilean is the sense in which most scientists use the word *experiment* today.

A Galilean is a *critical* experiment—one that discriminates between possibilities and, in doing so, either gives us confidence in the view we are taking or makes us think it in need of correction.

Galileo's having been born in Pisa made it inevitable that his superlative critical experiment on gravitational acceleration should be taken by everyone to have been executed by the dropping of cannonballs of different weights from the Leaning Tower. In reality, it was conducted without endangering life.

Galileo saw this kind of experiment as the ordeal *(il cimento)* to which we expose our hypotheses or the implications that follow from them.

Because of the asymmetry of proof explained below, experiments are very often designed not in such a way as to *prove* anything to be true—a hopeless endeavor—but rather to refute a "null hypothesis." As Karl Popper has pointed out, most general laws can be so construed as to *prohibit* the occurrence or deny the existence of certain phenomena or events. Thus the "law of biogenesis" declares that all living things are and always were the progeny of living things, and may thus be taken to prohibit the occurrence of spontaneous generation, the existence of which was made extremely doubtful by Louis Pasteur's brilliant experiments on bacterial putrefaction. Likewise, the

Second Law of Thermodynamics prohibits the occurrence of a great many phenomena that do not occur even in these permissive days. All prohibitions enforced by the Second Law are so many variants of the principle that speaks of the very extreme unlikelihood of passing spontaneously from a more probable to a less probable state. These prohibitions unfortunately include many plausible and profitable-sounding enterprises for designing self-energizing machines or machines of perpetual motion or for using twenty gallons of tepid bathwater to boil a kettle for one's coffee, and so on.

This possibility of casting many hypotheses into a negative form explains why so many experiments attempt to refute a null hypothesis—that which denies the validity of a hypothesis under investigation. The same principle applies to many statistical tests, where an example of R. A. Fisher's is as good as any: a tea drinker who professes always to be able to tell whether the milk has gone in first or last is exposed to *il cimento,* in which the null hypothesis is that her score of right and wrong guesses could perfectly well have been due to luck alone.

Although these various considerations can be spelled out logically, most scientists pick them up so quickly and so naturally that they seem almost instinctual in the way they go about their business. It is seldom said of any series of experiments that they "prove" the hypothesis under investigation, for long experience of human fallibility has taught scientists rather to say that their experimental findings or analyses "are (or are not) consistent with" the hypotheses under experimental investigation.

No experiment should be undertaken without a clear preconception of the forms its results *might* take; for unless a hypothesis restricts the total number of possible happenings or conjunctions of events in the universe, the experiment will yield no information whatsoever. If a hypothesis is totally permissive—if it is such that *anything* goes—then we are none the wiser. A *totally* permissive hypothesis says nothing.

The *"result"* of an experiment is never the *totality of observables;* the result of an experiment is almost always the *difference* between at least two sets of observables. In a simple, one-factor experiment, the two sets of observables are called the "experiment" and the "control." In the former, the factor under investigation is allowed to be present or to exercise its effects, and in the latter it is not. The "result" of the experiment is then the difference between the readings or counts in the experiment and the control. An experiment executed without a control is not Galilean in style but might still qualify as an experiment in the Baconian style—that is, as a little contrived performance of nature, though not a very informative one. In the performance of what is intended to be a critical experiment, clarity of design and fastidiousness of execution are the qualities to be aimed at.

It is a common failing—and one that I have myself suffered from—to fall in love with a hypothesis and to be unwilling to take no for an answer. A love affair with a pet hypothesis can waste years of precious time. There is very often no finally decisive yes, though quite often there can be a decisive no.

Kantian Experiments. Baconian, Aristotelian, and Galilean are not the only kinds of experiment. There are thought experiments, too; Kantian, I have called them in honor of the most breathtaking conceptual exploit in the history of philosophy: Kant's suggestion that instead of acquiescing in the ordinary opinion that our sensory intuitions are patterned by "objects"—by that which is perceived—we should take the view that the world of experience is patterned by the character of our faculties of sensory intuition. "This experiment succeeds as well as could be desired," Kant complacently remarked, and it led him to formulate his well-known opinion that *a priori* knowledge—knowledge independent of all experience—can exist; he reasoned that both space and time are *forms* of sensory intuition and as such are only "conditions of the existence of things as appearances." Before dismissing such an opinion as the merest metaphysical fancy, scientists should reflect that sensory physiology is becoming increasingly Kantian in tendency.[4] Another famous Kantian experiment is that which generates the classical non-Euclidean geometries (hyperbolic, elliptic) by replacing Euclid's axiom of parallels (or something

4. P. B. and J. S. Medawar, *The Life Science* (New York: Harper & Row, 1977), p. 147.

equivalent to it) with alternative forms. Demographic and economic projections are other examples of Kantian experimentation: "Let's see what would follow if we took a somewhat different view . . ."

Kantian experimentation requires no apparatus except sometimes a computer. The forms of experimentation characteristic of the natural sciences are Baconian and Galilean; upon these, it may be said, all natural science rests. In the historical, behavioral, and mainly observational sciences, exploratory activities normally end in the formulation of opinions of which the implications can be tested either by sociological field surveys, carbon dating, ascertaining the facts of the matter, referring to historical documents, or turning a telescope to a predetermined region of the sky. All such activities are Galilean in spirit—that is, they are critical evaluations of ideas.

The effect of Galilean experimentation is to preserve us from the philosophic indignity of persisting unnecessarily in error (the constant working of the process of rectification is discussed at length in Chapter 11). Any experienced scientist knows in his heart what a good experiment is: it is not just ingenious or well executed in point of technique; it is something rather sharp; a hypothesis does well to have stood up to it. Thus the merit of an experiment lies principally in its design and in the critical spirit in which it is carried out.

Elaborate and costly apparatus will sometimes be required, but no one should be taken in by the romantic notion that any scientist worthy of the name can carry out an experiment with no more apparatus than string, sealing wax, and a few empty bean cans; there is no conceivable method by which a sedimentation coefficient could be estimated with a bean can and string, unless someone is capable of swinging the can around his head more than a thousand times a second.[5] On the other hand, scientists must exercise discretion about the cost and complexity of the instruments they feel they need to use. Before commandeering costly plants and the services of colleagues night

5. The rotors of modern ultracentrifuges rotate at upwards of 60,000 times per minute.

and day, scientists should make very sure that their experiments are worth doing. It has been well said that "if an experiment is not worth doing, it is not worth doing well."

Discoveries

Experiments, then, are of many different kinds. So are discoveries. Some discoveries *look* as if they were merely a recognition or apprehension of the way nature is; they are lessons learned, as it were, by humbly taking note of what is going on; they have the air of being no more than "uncoveries" of what was there all the time, waiting to be taken note of. I myself believe it to be a fallacy that any discoveries are made in this way. I think that Pasteur and Fontenelle (see Chapter 11) would have agreed that the mind must already be on the right wavelength, another way of saying that all such discoveries begin as covert hypotheses—that is, as imaginative preconceptions or expectations about the nature of the world and never merely by passive assimilation of the evidence of the senses. It may of course be that an information-hunting exercise is that which prompts a hypothesis to take shape. Darwin's letters show that in believing himself to be a "true Baconian" he was simply deceiving himself.

Even so seemingly straightforward a discovery as that of a fossil is often the outcome of covert-hypothesis formation—for why otherwise should anyone look at the fossil remains twice and maybe take them back for more detailed investigation later? But how can we fit into this scheme such a remarkable discovery as that of the "living fossil" fish, the coelacanth *Latimeria?* What made this discovery so striking was this: most fossils —for example, those of the lungfish—are discovered *after* their living descendants have been recognized and described; it is most unusual for a fossil to be discovered before a living relative, as happened with *latimeria.* This is why its discovery gave the impression of a privileged and in some ways frightening insight into the world of very many million years ago.

Although I believe the same acts of mind underlie them both, I think it useful to draw a broad distinction between synthetic and analytic discoveries. A synthetic discovery is always a first recognition of an event, phenomenon, process, or state of affairs not previously recognized or known. Most of the stirring and deeply influential discoveries of science come under this heading. It is characteristic of a synthetic discovery that it need not have been made then and there—that it might, just conceivably, never have been made at all. Perhaps that is why we hold them in such awe.

My favorite example of this species is the discovery by Fred Griffith of the phenomenon of pneumococcal transformation,[6] which gave birth to modern molecular genetics. It turned out that the dead pneumococci that conferred some of their characteristics upon the living pneumococci in Griffith's famous experiment did not have to be whole and intact because extracts had the same effect. Some one particular chemical compound must have been responsible for the transformation. It was one of the great episodes in modern science when Avery, McLeod and McCarty showed this to be deoxyribonucleic acid (DNA). It in no way diminishes this discovery to describe it as "analytic" in character, for it was a triumph of intuition and experimental skill.

The character of an analytic discovery may also be illustrated by following the train of thought that led to the discovery of the structure of DNA. Ever since W. T. Astbury published the first X-ray crystallographs of DNA, imperfect though they were, it was recognized that DNA had a crystalline structure, probably of a repetitive or polymeric kind. The discovery of this structure was the outcome of the intellectual process described in Chapter 11—that is, the result of sustained dialogue between conjecture and refutation. But of course the distinction between synthetic and analytic is not hard and fast, for in the discovery of the structure of DNA there was both an analytic and a synthetic element, the latter being that its structure was just such as to equip it to encode and transmit genetic informa-

6. Pneumococcal transformation is a kind of transmutation of species that may occur when living pneumococci with one kind of carbohydrate capsule are mixed with dead pneumococci having capsules of another type. It sometimes happens that living organisms acquire some of the characteristics of the dead ones. See Medawar and Medawar, *The Life Science,* p. 88.

tion. This perhaps was the greater discovery, and in describing it as "the greater," I am speaking for the very widespread belief that synthetic discoveries—those which open up new worlds not until then known to exist—are those which scientists would most like to make.

But it would be wrong to make too much of discoveries. The greatest advances in modern biology have grown out of the intent and unrelaxing study of the characteristics of a single biological phenomenon or a single biological "system." This was the story of pneumococcal transformation and of protein synthesis in *Escherichia coli,* which showed the stages by which the structure of nucleic acid is mapped into the structure of a protein. So it will be now, I suspect, with the detailed mapping of the cell surface in respect of "histocompatibility" antigens. An individual discovery is here less important than the deep analysis that will eventually make known the molecular basis of specificity and help to explain why, in development, some cells go here rather than there, and some stick together though others do not. Deep analyses such as those of molecular biology will one day enable detailed molecular specifications to be drawn up for the synthesis of an enzyme or of an enzyme cascade that will, say, degrade polyethylene and thus reduce the proportion of the earth's surface occupied by the detritus of affluence.

For these reasons, a young scientist must not be disheartened if he does not become the eponym of a natural principle, phenomenon, or disease. Although the importance of discoveries may be overrated, no young scientist need think that he will gain a reputation or high preferment merely by compiling information—particularly information of the kind nobody really wants. But if he makes the world more easily understandable by any means—whether theoretical or experimental—he will earn his colleagues' gratitude and respect.

10

Prizes and Rewards

Scientists, like sportsmen and writers, are in the running for a whole variety of prizes and other rewards.

I knew a scientist who lost no opportunity to impress upon me his disapproval of the existence of such invidious distinctions, savoring as they did of elitism—the socially divisive notion that some people are better than others at some things—but when the opportunity came for his own nomination for the Fellowship of the Royal Society, he did not decline. Although in one of his Olympian moods that great mathematician G. H. Hardy referred to the Fellowship of the Royal Society as "a comparatively humble level of distinction," it is a greatly admired and eagerly sought-after reward for prowess in science. Ordinary membership is confined to British citizens, but honorary affiliations cast a wider net.

An elected F.R.S. is required to sign a book that contains the signatures of many of the greatest figures in the history of science; a new Fellow may indeed exult in being one of the company that includes Isaac Newton, Robert Boyle, Christopher Wren, Michael Faraday, Humphry Davy, James Clerk Maxwell, Benjamin Franklin and Josiah Willard Gibbs.

The Royal Society has a history going back to the days when a great revolution of the human spirit[1] inaugurated the modern

---

1. See Charles Webster, *The Great Instauration: Science, Medicine and Reform 1626–1660* (London: Butterworth, 1976).

world. It is far otherwise with the Nobel Prize, for the simple and sufficient reason that most of the very greatest scientists lived long before Alfred Nobel got the knack of stabilizing the nitric acid esters of polyhydric alcohols (especially glyceryltrinitrate) and founded the prize on the proceeds.[2] The Nobel award owes its great popular reputation to many things: public satisfaction in the expiatory element in the foundation of the award, the grand ceremony of the accolade, the size of the sum that changes hands, and the element of real distinction it embodies. But—and this is, I believe, the only valid ground for objecting to all such distinctions—all electoral procedures are fallible and the failure to gain a distinction of which a scientist is, and feels himself to be, genuinely worthy may cause not only great unhappiness but also personal injury to those whose livelihood and research support depends upon the judgment of people (for example, administrative high-ups) who may not realize how very many scientists are not, though they deserve to be, Fellows of the Royal Society of London or other comparable bodies. The same applies to the Nobel Prize, though it is difficult to feel the same sympathy because those who are not Nobel laureates but are sufficiently accomplished to be judged in the running are not likely to be embarrassed by lack of research funds.

Conventional wisdom has it that it is "bad" for the young to be successful too early: too many prizes and too high a scholastic record bode no good, we are sometimes told. "I'm afraid I wasn't very brainy at school," declares the pompous chump giving out the prizes, leaving us to infer that because of his other still more praiseworthy abilities it didn't handicap *him* a bit.

The supposed correlation between early success and later failure arises, I suspect, from one of those tricks of selective memory I have referred to elsewhere: of those who come to dust, it is the golden boys and girls we remember best; if they succeed—why, that was only to be expected, and so we remember only the failures.

---

2. Thus a world team (I. Newton, captain) of all-time great scientists chosen to meet a corresponding team from Mars or outer space could contain only a minority of Nobel Prize winners.

I have emphasized a darker side of prizes and rewards, but there is a very bright side, too: all such elections or nominations depend upon the good opinion that scientists are most eager for —the high opinion of their peers. The effect upon good scientists of gaining an award is a great moral boost—this expression of the confidence and esteem of others will promote their research and perhaps help them to do better than before. Very likely, too, the prizewinner will want to show everyone that it wasn't all a fluke.

In these respects, awards are wholly beneficent, but sometimes, unhappily, they have the opposite effect. I remember a fellow graduate student and I at Oxford telling each other in shocked voices of a university don who had said, "As soon as I get into the Royal, I shall give up research altogether." It seems only poetic justice that the occasion never arose for him to fulfill that ignoble ambition.

Of course, the head is sometimes turned by these distinctions, and there are Nobel laureates who give up research and spend their time traveling the world attending and sometimes addressing conferences with titles such as Science, Mankind, Values, and Human Endeavor (or any other such juxtaposition of abstract nouns). The vanity of such laureates is constantly inflamed by their being invited to sign and thus tip the scales in favor of the acceptance of some such manifesto as this: "The nations of the world must henceforward live together in amity and concord and abjure the use of warfare as a means of settling political disputes."

Can it be that a substantial number of people hold a contrary opinion but are suspending judgment until the signatures of fifty Nobel laureates convince them of its truth? It is all part of the human comedy, of course, but the exaggerated respect for prizewinners may sometimes be turned to useful ends—particularly in helping to secure the release from tyranny of prisoners of conscience, work in which Amnesty International has been particularly active.

It is fortunate that scientific honors cannot be worked for as one works for an exam; a young scientist can only hope that his work will be good enough for him to take his place one day among the candidature for such distinctions.

There is nothing ignoble about such an ambition, and that young scientists should cultivate it was often a principal purpose of the founders or sponsors of the award.

11

The Scientific Process

Je cherche à comprendre
—Jacques Monod

How do scientists go about making discoveries, propounding "laws," or otherwise enlarging human understanding? The conventional answer, "by observation and experiment," is certainly not mistaken, but it needs to be interpreted with reserve. Observation is not a passive imbibition of sensory information, a mere transcription of the evidence of the senses, and experimentation is not only of the kind that I classified as Baconian in Chapter 9—that is, the contrivance of phenomena or conjunctions of events that do not occur spontaneously in nature. Observation is a critical and purposive process; there is a scientific reason for making one observation rather than another. What a scientist observes is always a small part only of the whole domain of possible objects of observation. Experimentation, too, is a critical process, one that discriminates between possibilities and gives direction to further thought.

A young scientist has now a meter or so of bench space, let us say, a white coat, authority to use the library, and a problem that he has thought up himself or that a senior has asked him to look into. To begin with, anyway, it is almost certain to be a small problem—one of which the solution will facilitate the solution of something more important, and so on, until the long-term objective of the enterprise is in sight. Nonscientists cannot immediately see the connection between the lesser problem and the greater. It must often occur to a humanist as he reads the minutes of the board of the faculty of science that young scientists are engaged in comically specialized activities. A scientist might equally well wonder what there could be to engage a grown man in the study of the parochial affairs of Tudor Cornwall, because he does not realize that such an investigation is about the Reformation, a very great affair indeed.

But what will a scientist *do* to resolve his problem? Something of which he can be quite certain is that no mere compilation of factual information will serve his purpose.[1] No new truth will declare itself from inside a heap of facts. It is true that Bacon and Comenius and Condorcet too (see below) sometimes wrote as if they believed that the collection and classification of empirical facts would lead to an understanding of nature, but in taking this view they were guided by a rather special consideration: they felt under a strong obligation to refute the idea that *deduction* was an act of mind that could lead to the discovery of new truths—that an act of mind alone could enlarge the understanding. The philosophic and scientific writing of the seventeenth century—particularly the writing of Bacon, Boyle, and Glanvill, for example—is full of dismissive references to Aristotle's way of thinking, in the tradition of which they had all grown up.

Bacon's exhortation to observe and to experiment does not, of course, tell the whole story of his scientific philosophy; he also propounded a number of rules for getting at the truth of things essentially similar to those which two hundred years later John Stuart Mill propounded as the rules of discovery in his *System of Logic.* These rules of induction are applicable only under special circumstances: when we have before us all and only the facts relevant to the solution of our problem—the whole truth and nothing but. Thus we may be called upon to conduct an epidemiological exercise to account for the violent sickness of

1. To avoid repeated acknowledgments of indebtedness, I mention here once for all that the account of the scientific process that follows is based very largely upon the writings of Sir Karl Popper, F.R.S., especially *The Logic of Scientific Discovery,* 3d ed. (London: Hutchinson, 1972) and *Conjectures and Refutations,* 4th ed. (London: Routledge & Kegan Paul, 1972).

a member of a dinner party; we know what they all ate and drank, we know that all were hale when they sat at table, and that all but the victim remained so afterward. On this basis, the so-called rules of induction can be applied; the things eaten by everyone are not likely to be responsible for the illness of only one, nor is the dish that everyone refused: only the victim, it turns out, ate the cream syllabub. Only a singularity of exposure to risk can account for the victim's singular misfortune. These simple exercises in elementary logic and common sense are hardly worth dignifying by the long appellations Bacon gave them. The rationale of fact-hunting in the eyes of such as Mill and Bacon was that it would put the scientist in possession of the facts upon which such a calculus of discovery could be made to work.

In real life it is not like this. The truth is *not* in nature waiting to declare itself, and we cannot know *a priori* which observations are relevant and which are not; every discovery, every enlargement of the understanding begins as an imaginative preconception of what the truth might be. This imaginative preconception—a "hypothesis"—arises by a process as easy or as difficult to understand as any other creative act of mind; it is a brainwave, an inspired guess, the product of a blaze of insight. It comes, anyway, from within and cannot be arrived at by the exercise of any known calculus of discovery. A hypothesis is a sort of draft law about what the world—or some particularly interesting aspect of it—may be like; or in a wider sense it may be a mechanical invention, a solid or embodied hypothesis of which performance is the test.

Thus the day-to-day business of science consists not in hunting for facts but in testing hypotheses—that is, ascertaining if they or their logical implications are statements about real life or, if inventions, to see whether or not they work. In the Galilean sense (see Chapter 9) in which I said the word *experiment* is now most widely used, experiments are the acts undertaken to test a hypothesis.

In the outcome, science is a logically connected network of theories that represents our current opinion about what the natural world is like.

Once he has a hypothesis to work on, the scientist is in

business; the hypothesis will guide him to make some observations rather than others and will suggest experiments that might not otherwise have been performed. Scientists soon pick up by experience the characteristics that make a good hypothesis; as explained in Chapter 9, almost all laws and hypotheses can be read in such a way as to *prohibit* the occurrence of certain phenomena (the example I gave was the prohibition by the law of biogenesis of the occurrence of spontaneous generation). Clearly, a hypothesis so permissive as to accommodate *any* phenomenon tells us precisely nothing; the more phenomena it prohibits, the more informative it is.

Again, a good hypothesis must also have the character of *logical immediacy,* by which I mean that it must be rather specially an explanation of whatever it is that needs to be explained and not an explanation of a great many other phenomena besides. It is not wrong but equally it is not very helpful to interpret Addison's disease or cretinism as the consequence of a "malfunction of the hormone-secreting glands." The great virtue of logical immediacy in a hypothesis is that it can be tested by comparatively direct and practicable means—that is, without the foundation of a new research institute or by making a journey into outer space. A large part of the *art of the soluble* is the art of devising hypotheses that can be tested by practicable experiments.

Most of the everyday business of the empirical sciences consists in testing experimentally the logical implications of hypotheses—that is, the consequences of assuming for the time being that they are true. The experiments I described as critical or Galilean give direction to further speculation: their results either square with the hypothesis under consideration, in which case it remains on probation while some further and more searching tests are planned, or else cause the hypothesis to be revised or in the extreme case to be abandoned altogether, whereupon the dialogue must being anew. The dialogue I envisage is between the possible and the actual, between what *might* be true and what is in fact the case—a dialogue between two voices, the one imaginative and the other critical, between *conjecture and refutation,* as Popper has it.

These acts of mind are characteristic of *all* exploratory processes and are certainly not confined to experimental sciences, for this is essentially how an anthropologist will proceed, a sociologist, or a physician intent upon diagnosis. It is also the process of mind used by the mechanic who tries to figure out what is wrong with a car. It is all very far removed from the fact-hunting of classical inductivism. As a point of logic that has some bearing on the way he thinks he goes about his business, a young scientist must always avoid saying or thinking that he "deduces" or "infers" hypotheses. On the contrary, a hypothesis is that *from* which we deduce or infer statements about matters of fact, so that, as the great American philosopher C. S. Peirce clearly recognized, the process by which we try to think up the hypotheses from which our observations will follow is an inverse form of deduction—a process for which he coined the terms *retroduction* and *abduction,* neither of which has caught on.

Some Implications of These Views

Feedback. Although it has been pointed out very often, there is no harm in pointing out again that if the inferences we draw from a hypothesis are thought of as its logical output, then the process by which we modify a hypothesis in accordance with the degree of correspondence of its predictions to reality is yet another example of the fundamental and ubiquitous stratagem of negative feedback (see "Falsification," below). This parallel reminds us that scientific research, like other forms of exploration, is, after all, a cybernetic—a steering—process, a means by which we find our way about, and try to make sense of, a bewildering and complex world.

Falsification and the Asymmetry of Proof. The recognition of the asymmetry of proof is fundamental to an understanding of the scheme of thought just outlined (the "hypothetico-deductive" scheme).

Consider a simple syllogism from schoolroom logic:

major premise: All men are mortal.
minor premise: Socrates is a man.
inference: Socrates is mortal.

If correctly executed, the process of deduction brings with it the complete and unqualified assurance that if the premises are true, then the inference must also be true. Socrates must indeed be mortal. No question. But this is a one-way process; the mortality of Socrates, supposing that historical research confirms it, gives us no positive assurance of his having been a man or of the mortality of mankind generally. The syllogism and the inference would be equally binding upon us if Socrates were a fish and all fish mortal. We can, however, say with complete certainty that if Socrates were *not* mortal—that is, if the inference were wrong—then we must be thinking on the wrong lines: either Socrates was not a man or not all men are mortal.

The upshot of this asymmetry of inference is that falsification is a logically stronger process than what sometimes people rather recklessly refer to as "proof"; indeed, a scientist does not very often speak with complete confidence of "proof." The more experienced he is, the less likely he is to do so. As they grow in experience, scientists soon come to appreciate the special strength of falsification and the precariousness of what beginners call "proof," for as explained in Chapter 9 (where a different reason for this experimental design was given), it is a well-known stratagem of research to investigate and mayhap refute the "null" hypothesis, which affirms the very opposite of whatever may be under investigation. For all these reasons no hypothesis in science and no scientific theory ever achieves apodictic certainty—never achieves a degree of certainty beyond the reach of criticism or the possibility of modification.

A scientist is, then, a *seeker after truth.* The truth is that which he reaches out for, the direction toward which his face is turned. Complete certainty is beyond his reach, though, and many questions to which he would like answers lie outside the universe of discourse of natural science. The last words of one of the greatest scientists of the twentieth century, Jacques Lucien Monod, which I have used as the motto of this chapter, embody an ambition that a scientist can always achieve: he can try to understand.

What Is a Scientific Statement? Scientists who in their professional capacities make scientific statements may sometimes be too ready to accuse others of being "unscientific," so it would be useful to have a criterion, a line of demarcation to make it possible to distinguish between statements that belong to the world of science and of common sense and those that belong to some other world of discourse.

When logical positivists first tackled this problem, they felt they had the answer in the notion of "verification." Scientific statements were verifiable in fact or in principle; verifiability "in principle" was enjoyed by those statements of which it was possible to see what steps should be or could be undertaken to verify them. Statements not verifiable in principle were dismissed as "metaphysical"—a word clearly used as a euphemism for nonsense. Karl Popper, because of his special and well-founded views on the efficacy of falsification, substituted "falsifiability in principle" for "verifiability in principle." The new line of demarcation he proposed was *not,* he insisted, between sense and nonsense, but simply between two different worlds of discourse, the one belonging to the world of science and common sense, the other to metaphysics and serving altogether different purposes.

Where Does Luck Come into All This? "Serendip" was an old-fashioned name for Ceylon. It was a conceit of Horace Walpole's that the three princes of Serendip were forever coming upon felicitous discoveries or inventions by good luck alone: hence, "serendipity."

Luck plays a real part in scientific research, and after long periods of discouragement or following pathways of research that lead nowhere, scientists often say or think they are about due for a lucky break. By this they do not mean anything that would be judged lucky by the criterion of induction—a lucky presentation to their senses, ready-made, of some important new phenomenon or conjunction of events. What they mean is that it's about time they had a right idea instead of a wrong one—about time they hit upon a hypothesis that not only ostensibly explains what is to be explained but also stands up to critical evaluation.

Dr. Roger Short has given a most interesting example of the

inadequacy of mere observation in discovery. It gains special force from the fact that William Harvey was a superlative *observer.* Writing of Harvey's conception of conception, Short points out that he dismissed altogether the complicity of the ovaries in mammalian reproduction, believing with Aristotle that the egg was a product of conception and especially of the male "seed." Short adds: "Harvey's dissections and observations were almost faultless, and it was only in their interpretation that he erred. His mistake may even serve as a lesson to many of us today."[2]

But what about luck in a more familiar and less intellectual sense? What about, for example, the discovery by Alexander Fleming of penicillin?

Fleming was a fine scientist and therefore not too grand to set up his own bacterial culture plates. The myth (for so I have been told it is), however, goes as follows. One day, when Fleming was setting up a plate of staphylococci or streptococci, a spore of the bread mold *penicillium* floated in through the window and settled upon his culture plate. Around the spore there developed a halo of inhibition of bacterial growth, the germinal discovery from which all the rest followed.

For very many years I accepted this story because I had no reason or inclination to do otherwise, but a cynical bacteriologist at the British Postgraduate Medical School in Hammersmith challenged it on several grounds. First of all, a spore of penicillin will not germinate in this way to give rise to a zone of inhibition of bacterial growth. The bacteriologist went on to tell me that St. Mary's was an old-fashioned building, the windows of which would either not shut or not open. Fleming's were of the latter kind; so much for the spore's floating in through the window.

I was sorry that the traditional story of Fleming's discovery did not stand up to critical scrutiny because I should have liked to have believed it true; but even if it had been true, it would not have told us very much about the efficacy of luck. Fleming

2. R. V. Short, "Harvey's Conception," in *Proceedings of the Physiological Society* (July 14–15, 1978). See also R. V. Short, in Zuckerman, ed., *The Ovary,* vol. 1, 2d. ed. (New York: Academic Press, 1977).

was a humane and gentle man who had been shocked and sickened by the gangrene and other horrible complications he had found in the battle casualties of World War I. The phenolic antiseptics that alone were available were almost completely inactivated by body fluids and would have damaged the tissues of the body more than the bacteria, thus adding to the complications of an infected wound. Fleming therefore had clearly in mind the special advantages of an antibacterial substance that did not damage tissues.

It is not methodologically an exaggeration to say that Fleming eventually found penicillin because he had been looking for it. A thousand people might have observed whatever it was that he did observe without making anything of it or building upon the observation in any way; but Fleming had the right slot in his mind, waiting for it. Good luck is almost always preceded by an expectation that it will gratify. Pasteur is well known to have said that fortune favors the prepared mind, and Fontenelle observed, *"Ces hasards ne sont que pour ceux qui jouent bien!"* ("These strokes of good fortune are only for those who play well!").

There *was* one amazing stroke of pure good luck about penicillin for which no one's mind could possibly have been prepared because only recent research has brought it to light: most antibiotics are exceedingly toxic because they interfere with a department of bacterial metabolism shared by bacteria and ordinary body cells. Actinomycin D provides a good example because it interferes with the mapping of the DNA of the cell nucleus into the RNA through which its genetic effects are exercised; because the mechanism is common to both, actinomycin affects ordinary body cells as it does bacteria. Penicillin is not toxic because it affects metabolism of a kind peculiar to bacteria.

Limitations of Science. If we accept, as I fear we must, that science cannot answer questions about first and last things or about purposes, there is yet no known or conceivable limit to its power to answer questions of the kind science *can* answer. The founding fathers of the seventeenth century were not mistaken in taking *plus ultra* as a slogan—in believing that in science there is always more beyond. When Whewell first propounded a view of science of the same general kind as that which Karl Popper has developed into a thoroughgoing system, his opponent John Stuart Mill was shocked by the reflection that hypotheses were products of the imagination and had no confinements, therefore, other than those of the imagination itself; yet what scared Mill is one of the great glories of science and our principal assurance that it has no limit. Science will dry up only if scientists lose or fail to exercise the power or incentive to imagine what the truth might be. One can envisage an end of science no more readily than one can envisage an end of imaginative literature or the fine arts. Some problems may be insoluble, of course; Karl Popper and John Eccles have commented that the connection between brain and mind might be one,[3] but it is not easy to think of a second.

The March of Paradigms

My partiality for the "hypothetico-deductive" account of the scientific process has been based on as accurate a study as I have found it possible to carry out on my own processes of thought, abetted by opinions of the fairly large number of scientists and physicians who have come to think it a fair representation of the exploratory process; but it would be very unfair to create the impression that the scheme I have outlined is the only prevailing interpretation of the scientific process. Great interest was aroused by the views expounded by Thomas Kuhn in *The Structure of Scientific Revolutions* and more recently in *Essential Tension.*[4] There is an illuminating discussion of Kuhn's view by Kuhn himself and others in a symposium entitled *Criticism and the Growth of Knowledge.*[5]

Kuhn's views have caught on—a sure sign that scientists find them illuminating because they haven't much time for what

---

3. Karl R. Popper and John C. Eccles, *The Self and Its Brain* (Berlin: Springer, 1978), Preface.

4. Thomas Kuhn, *The Structure of Scientific Revolutions* (Chicago: University Press, 1962; 2d ed., 1970); *Essential Tension* (Chicago, Ill.: University Press, 1978).

5. I. Latakos and A. Musgrave, eds., *Criticism and the Growth of Knowledge* (Cambridge: Cambridge University Press, 1970).

they think of as mere philosophizing. Kuhn's views and Popper's are not antithetical.

Kuhn's position is in outline this. In the critical evaluation of hypotheses to which Popper rightly attaches such great importance, the evaluation of a hypothesis is not a private transaction between the scientist and reality—a competition, as it were, between fact and fancy. That which the scientist measures his hypotheses against is the current "establishment" of scientific opinion—the current framework of theoretical commitments and received beliefs—the prevailing "paradigm" in terms of which the day-to-day problems arising in a science tend to be interpreted. A scientist who explores within its ambience is executing what Kuhn calls "Normal Science," and his researches are so much puzzle-solving.

It is no wonder that J. W. N. Watkins in the symposium to which I have referred above remarked that Kuhn sees the scientific community on the analogy of a religious community, with a science as a scientist's religion. It is true, certainly, that scientists are often reluctant to shake off received beliefs and sometimes feel impatient of notions that fall outside the prevailing paradigm, but normal science does not long persist unchallenged; every so often, an extraordinary scientist or extraordinary scientific phenomena supplant the prevailing paradigm by a new orthodoxy—a new paradigm that defines a "normal" science anew and lasts until the revolutionary appraisal is repeated. The "essential tension" to which Kuhn refers in the title of his latest book is between our inheritance of doctrine and dogma as they affect science and the occasional upheavals that inaugurate a new "paradigm" in the terminology Kuhn has made popular.

Kuhn's views throw some light on the psychology of scientists and are an interesting comment on the history of science, but they do not add up to a methodology—a system of canons of inquiry.

In real life, a scientist tends to believe in a hypothesis until he has reason to do otherwise. This, then, is his personal paradigm, reinforced perhaps by some pride of possession if it embodies an idea of his own. As for revolutions, they are constantly in progress; a scientist does not hold exactly the same opinions about his research from one day to the next, for reading, reflection, and discussions with colleagues cause a change of emphasis here or there and possibly even a radical reappraisal of his way of thinking. In a laboratory there are continual movements of unrest. There is something about Kuhn's writing that makes me think that he sees normal scientific life as one of settled, God-fearing bourgeois contentment within an established order of things, but in reality it is more like a Maoist microcosm of continuing revolution; in any laboratory conducting original investigation, all is in flux. It may, of course, be different in the social sciences, which have a slower pulse and in which an opinion takes very much longer to appraise. Here perhaps we may speak of a "normal science," and the process by which it is supplanted may be likened more aptly to a revolution.

Is There Too Much Fuss About Method? Even though an episode of scientific inquiry can be shown in retrospect to have a hypothetico-deductive character, a young scientist may well wonder if there need be any great formality about it all; most scientists, he may reflect, have received no formal instruction in scientific method, and those who have seem to do no better than those who have not.

A young scientist has no need to exercise a methodology in any highfalutin sense; he must realize very clearly, though, that collecting facts could at best be only a kind of indoor pastime. There is no formulary of thought or program of ratiocination that can conduct him quickly from empirical observations to the truth. An act of mind always interposes between any observation and any interpretation of it. The generative act in science, I have explained, is imaginative guesswork. The day-to-day business of science involves the exercise of common sense supported by a strong understanding, though not using anything more subtle or profound in the way of deduction than will be used anyway in everyday life, something that includes the ability to grasp implications and to discern parallels, combined with a resolute determination not to be deceived either by the evidence of experiments poorly done or by the attractiveness —even lovableness—of a favorite hypothesis. Heroic feats of intellection are seldom needed. "The scientific method," as it is sometimes called, is a potentiation of common sense.

Before he sets out to convince others of his observations or opinions, a scientist must first convince himself. Let this not be too easily achieved; it is better by far to have the reputation for being querulous and unwilling to be convinced than to give reason to be thought gullible. If a scientist asks a colleague's candid criticism of his work, give him the credit for meaning what he says. It is no kindness to a colleague—indeed, it might be the act of an enemy—to assure a scientist that his work is clear and convincing and that his opinions are really coherent when the experiments that profess to uphold them are slovenly in design and not well done. More generally, criticism is the most powerful weapon in any methodology of science; it is the scientist's only assurance that he need not persist in error. All experimentation is criticism. If an experiment does not hold out the possibility of causing one to revise one's views, it is hard to see why it should be done at all.

12

Scientific Meliorism Versus Scientific Messianism

Scientists are characteristically sanguine in temperament, a state of mind sometimes thought to contrast rather discreditably with that which Stephen Graubard has called the "habitual despondency of the literary humanists." It is not to be wondered at, though, having regard to the fact that in terms of the fulfillment of declared intentions science is incomparably the most successful activity human beings have ever engaged upon, though we don't hear much about the airplanes that did not fly, and most discarded hypotheses are secret sorrows.

Sanguine though scientists may be, it would be a philosophic error to describe them as "optimists," for if they were so, much of their *raison d'être* would disappear. Optimism, a metaphysical belief growing out of Leibnitz's theodicy, did not survive Voltaire's ridicule; Voltaire's *Candide* did it in. All is *not* well, his message runs; this is not the best of possible worlds.

Utopia and Arcadia

Scientists tend also to be Utopian in temperament—to believe in the possibility in principle, perhaps even in fact, of a different and altogether better world. The great days of Utopian thinking were the days when voyages of discovery on the earth's surface had the same significance as space travel has today. The old Utopias—New Atlantis, Christianopolis and the

City of the Sun—were faraway contemporary societies, but the Utopias men dream of today lie in the distant future or on a planet of a distant undiscovered sun.

Arcadian thinking looks not forward nor far away but backward to a golden age that could yet return. Arcadia is a world of innocence not yet corrupted by ambition and inquiry, a world of pious acquiescence in the established order of things, without strife and without ambition—a world of "truth and honest living." Milton, whom I quote, saw it as the purpose of education "to repair the ruins of our first parents," to return to the happy innocence of the world before the Fall. Arcadian ambitions were not uncommon in the millenarian beliefs of Puritan intellectuals contemporary with Milton. We need not wonder that they played—as Charles Webster has so clearly shown in *The Great Instauration: Science, Medicine and Reform 1626–1660*[1]—a very important part in the scientific revolution of Bacon and Comenius, for both their Arcadian beliefs and their championship of the new philosophy were manifestations of an extreme dissatisfaction with the world as it had become.

Arcadian thinking is not dead today; it simply takes a different form. Although the notion of a cyclical recurrence of historic epochs has been abandoned, it is motivated still by dissatisfaction—especially with the world for which, as it is believed, "science is responsible."

One such latter-day Arcadia envisages as the highest condition of man the state of the prosperous English landed gentleman of the eighteenth century. Living on the wholesome and abundant produce of the home farm, he was surrounded by a contented and respectful tenantry, whose interests he genuinely looked after; he gave employment, moreover, to a large number of loyal indoor and outdoor servants, to whom his convocation for morning prayers or regular attendance at church set an example of manly piety. The landed gentleman raised a

1. Charles Webster, *The Great Instauration: Science, Medicine and Reform 1626–1660* (London: Butterworth, 1976). The quotation from Milton is from his letter on education to Samuel Hartlib (1644), reprinted in the Everyman edition of Milton's prose works.

large family, the eldest male member of which would succeed him in the care and management of his estate; his daughters, when not supporting their mother in the execution of all manner of good work, added still further distinction to the family name by advantageous marriages. To complete the Arcadian microcosm, a young resident tutor, with an eye perhaps on the family living, did his best to educate the young in a style Dr. Johnson would have approved (see page 33).

This was without doubt a wonderful world for the landed gent himself, but nothing like so much fun for the domestic staff, none earlier to bed than the latest sitter-up and some up at dawn to lay fires in the bedrooms and living rooms and have everything shipshape before the quality came down. The outdoor staff worked very hard, too, probably deriving less satisfaction than their masters from contemplating their place in the established order of things, being at all times conscious that their own and their families' livelihoods depended on the approval and goodwill of the landed gentleman or his agent.

Nor was it so much fun for the landed gentleman's wife; she strove by repeated childbearing to make good the ravages of a merciless infant mortality and might be condemned to nurse—in secret—painful and disabling ailments that pride, propriety, and a well-founded doubt of the efficacy of medical treatment made it useless to declare. Her own bondage to the established order of things was no less absolute and perhaps in some ways more demanding than that of the domestic staff.

C. S. Lewis, from whom in friendly conversations lasting several years I reconstructed the more agreeable elements of this Arcadian dreamland, had it always in mind as a counterirritant to the science-based world that he abhorred. He thought scientists were plotting to supplant the world he loved best with the produce of factory farms and chemical agriculture—a barren world indeed, as he saw it: "no high chairs, not a gleam of gold, not a hawk, not a hound," he wrote in *That Hideous Strength*—but of course Lewis saw himself as the landed gent, as all do who indulge themselves in this Arcadian fancy. Scientists seldom have the upbringing and the worldly wisdom to cast themselves as principals and would be more likely to wonder what it would be like to be at best the resident tutor or, more likely, the chap who went around unstopping the drains.

The Arcadia I have just outlined is of course fairly recent; it is very far removed from that primitivism which found its best-known expression in the noble savage of Jean Jacques Rousseau. Long before Rousseau there had been speculation about a world of primitive innocence and plenty—for example, about a Hyperborean community living where the earth poured forth its bounty and goats came of their own accord to be milked.

This primitivism has been an important element in human cultural history, and so far from obliterating it, the growth of science has made it even more attractive if less plausible than before. Anyone on the lookout for it will find in everyday life and thought plenty of evidence of how often Rousseau rides again.

Scientific Messianism

Sanguine or despondent, Utopian or Arcadian in temperament, scientists, like most other folk, want to feel they have some special reason for being alive—not just for "being in this world," as the saying is, but for being a scientist rather than anything else.

One soon picks up from the conversations or declared opinions of scientists, especially young ones, that the belief that animates many of them is what Sir Ernst Gombrich has called "scientific messianism." It goes naturally with Utopianism—a better world is possible in principle and may be brought into existence by a great transformation of society. Science, they believe, will be the agent of this transformation, and the problems that beset mankind—not excluding those which grow out of the imperfections of human nature—will yield to a scientific inquiry that will point the way to those sunlit uplands of peace and plenty that seem like a secular heaven to a weary and rather battered world.

This great and deep faith in science rose out of two great revolutions of the human spirit. The first, of which Francis Bacon was evangelist, ushered in the new philosophy ("new science," we now say). Bacon's *New Atlantis* was his dream of what a world shaped by this new philosophy might be: a world of which the principal commodity was light—the light of understanding, not only of the material world but also of our fellow creatures. The philosopher-scientists who governed this world were dedicated to *the effecting of all things possible* through the indefinite enlargement of human understanding.

Nothing now remains of Bacon's Atlantidean dream except that element which embodies both the glory and the threat of science: the conscious recognition of the truth that everything that is in principle possible—which does not contravene a natural law—can be done if the intention to do it is sufficiently resolute and sufficiently long-sustained. It is a corollary of this truth that the direction of scientific endeavor is determined by political decisions, or at all events by acts of judgment that lie outside science itself. Science opens up possible pathways of action but does not itself point to one rather than another.

Charles Webster, to whose great work on the world of Bacon and Comenius I have already referred, pointed out that much of the motivation of the new philosophy came from radical Puritan activists who saw in the new science the means of making England fit to be host nation to the impending millennium, fulfilling the prophecy embodied in Daniel 12:4: "Many shall run to and fro and knowledge shall be increased." It was not by chance that the 1620 edition of Bacon's *Great Instauration* showed ships passing freely through the Straits of Gibraltar, at one time thought to mark the limit of the world. Vast seas are visible beyond the Pillars of Hercules, for there was always *plus ultra*—more beyond. "Come, come, come," wrote Samuel Hartlib in a letter to Jan Amos Comenius, exhorting him to come to England, "it is time for the servants of the Lord to gather in one place and prepare the table for the coming of the Lord's anointed." The development of the sciences and the useful arts was to be a most important element in the preparation of the table.

It is a principal lesson of Webster's treatise—surprising to those brought up in more conventional views—that modern science has deeper religious and literally scriptural origins than are generally recognized. The period chosen by Webster for special investigation, 1626–60, was intellectually the most exciting and exhilarating period in the modern world, an era of great hopes and beginnings; science was then dominated by men in holy orders whose professional advancement depended to a large extent on Puritan patronage.

Although Bacon described himself as the "trumpeter" of the new philosophy, very much of his thinking had a medieval or still more ancient cast (Professor Paulo Rossi called him "a medieval philosopher haunted by a modern dream"), and although his scientific method did not, and indeed could not, work, Bacon's writing fired and inspired his readers, and can do the same today. He is still science's greatest spokesman, still the greatest evangelist; we can still recapture through the writings of Bacon and Comenius the exultation and breathless excitement that went with the inauguration of the world we now live in.

The second great movement of thought that helped to create the messianic conception of science was marked not so much by exhilaration as by truly wonderful complacency and self-confidence. It was that which we call the Enlightenment. For Condorcet, its most touchingly dedicated spokesman, progress had an historical inevitability. The present state of mankind in "the most enlightened countries of Europe" was such, he said, that philosophy (science) "has no longer anything to guess, has no more suppositious combinations to form; all it has to do is to collect and arrange facts, and exhibit the useful truths which arise from them as a whole, and from the different bearings of their several parts." Progress, he believed, was assured by the constancy of the laws of nature; Condorcet accordingly undertook to show how this progress, "chimerical though it might appear to be, was gradually to be rendered possible and even easy," and how "truth, in spite of the transient success of prejudices, and the support they receive from the corruption of governments or of the people, must in the end obtain a durable triumph." Nature, he went on to explain, had "indissolubly united the advancement of knowledge with the progress of liberty, virtue, and respect for the natural rights of man."

It takes the breath away still, this calm assurance of the inevitability of progress mediated through scientific learning. A man of Condorcet's hopeful innocence could not—did not—avoid the enmity of revolutionaries. The work from which I quote (in a contemporary translation) was published after his death at their hands.

Scientists as a class are rationalists, at least in the limited sense of believing without qualification in the *necessity* of reason. They would be surprised and offended if any withdrawal from such a view were imputed to them. Rationalism carries with it a professional obligation to combat the modern taste for irrationalism—not just spoon-bending (a fashionable form of psychokinesis) or its philosophic equivalents, but the inclination to substitute "rhapsodic" intellection for the humdrum ratiocination that has satisfied all the world's great thinkers hitherto. Among the principal antiscientific movements are the cult of the wisdom of the East and of mystical theology—a prose offering to the Almighty, said George Campbell, which, where a living sacrifice would have been deprived of life, had been deprived, instead, of—sense.

Young scientists must however never be tempted into mistaking the necessity of reason for the sufficiency of reason. Rationalism falls short of answering the many simple and childlike questions people like to ask: questions about origins and purposes such as are often contemptuously dismissed as nonquestions or pseudoquestions, although people understand them clearly enough and long to have the answers. These are intellectual pains that rationalists—like bad physicians confronted by ailments they cannot diagnose or cure—are apt to dismiss as "imagination." It is not to rationalism that we look for answers to these simple questions because rationalism chides the endeavor to look at all.

Scientific Materialism Examined

A scientist who works for the advancement of medicine or agriculture or the improvement of manufactures can be—often is—an agent of material progress. As such, he will be frowned upon for two different reasons: the first is that which is embodied in the well-known cliché of second-rate criticism to the effect that material prosperity entails spiritual impoverishment; and the second, much more serious, is that material progress does not hold out the promise of remedying any of the major ailments that afflict mankind today.

The idea that material prosperity entails spiritual impoverishment is a favorite of those who deride the idea of progress, though many who do so—or who lose sight of the notion altogether in a simulated agony of puzzlement about what progress "really" means—are secretly believers; very few people genuinely prefer bad drains to good, although as Bryan Magee has pointed out,[2] *The Times* of London could at one time have been counted among the former. *The Times* rapped Edwin Chadwick sharply over the knuckles for presuming to improve the health of Londoners by laying down adequate sewers. No, declared *The Times,* speaking with the voice of antiscience throughout the ages, Londoners would rather "take their chance with cholera and the rest than be bullied into health by Mr. Chadwick and his colleagues." Ironically, because he was a great believer in progress, Albert the Prince Consort was one who had to take his chance. When he died of typhoid, the twenty cesspools in Windsor Castle were found to be full to overflowing.

The spirit of *The Times*'s denunciation of Edwin Chadwick is still abroad; every time the mayor of an American municipality finds against fluoridation or someone in England pronounces it inefficacious or even downright harmful, there is a clamor of rejoicing in the corner of Mount Olympus presided over by Gaptooth, the God of Dental Decay.

Once again we are obliged to draw a distinction between the sufficient and the necessary. For the full unfolding of the human spirit good drains, speedy communications and sound teeth are not sufficient, but they help. There is nothing about poverty, privation, and disease that is conducive to creativity; let no one be taken in by such romantic nonsense. Florence in its greatest days was a great mercantile and banking center; Tudor England was a bustling and prosperous country; and we may look in vain to Rembrandt's Amsterdam for evidence that the arts flourish in adversity. Although I do not often hear remarks of

2. Bryan Magee, *Towards Two Thousand* (London: MacDonald, 1965).

such surpassing idiocy, I can remember being assured that Switzerland was a good example of a country in which prosperity and material comforts, whether the product of science and industry or of frugality and good housekeeping, had stifled the creative afflatus for good.

Switzerland's principal contribution to civilized life, the knowing voice continued, was the cuckoo clock. It is an amazing judgment that attaches no importance to the lesson Switzerland has taught the world of peaceful coexistence in a multinational community, of the tolerance and hospitality that have long made it a retreat for philosophers, scientists, imaginative writers, and fugitives from tyranny.

The real case against the adequacy of the material progress made possible by science rests upon the exposure of a simple doctrinal fallacy that is a modern secular equivalent of the doctrine of original sin: the *doctrine of original virtue.* Give human beings the assurance of food, warmth, shelter and freedom from pain, and their natural goodness will prevail—they will become peaceable, loving, and cooperative, eager to help others and to work for the common weal. Give children love and warmth and protection, and they will be loving and lovable, outgoing and unselfish, sharing their toys and other possessions spontaneously with their friends, enjoying a clear instinctual perception of what is best for them at the time and thereafter. Inexperienced teachers and young parents do sometimes seriously believe that children not only know what is best to eat and best to do but also what they should learn or not learn; they also have serious misgivings lest firmness or an exercise of authority should deprive children of their spontaneous creativity and innocent perceptiveness.

Nothing, I think, has ever formally disproved the doctrine of original virtue, though there is precious little incentive to believe it true. Yet one cannot help thinking the tendency to believe in it is a lovable human trait.

Scientific Meliorism: A Realistic Ambition for Science

If the doctrine of original virtue were true, then scientific messianism would embody a valid ambition because science could one day create the ambience in which natural virtue would prevail; but let us consider instead what lesser ambition scientists might entertain for science.

Many young scientists hope that the science they come to love can be the agent of a social transformation leading to the betterment of mankind; accordingly they lament that so few politicians are scientifically trained and that so few have a deep understanding of the promise and the accomplishments of science. These lamentations betray a deep misunderstanding of the nature of the most exigent problems that confront the world: the problems of overpopulation and of achieving harmonious coexistence in a multiracial society. These are not scientific problems and do not admit of scientific solutions. This does not mean that scientists are confined to being shocked spectators of events or political dispositions that threaten the well-being of nations and ultimately mankind; scientists, as scientists, will find that they have necessary and distinctive contributions to make to the solution of these problems—but they are solutions that fall short of ushering in the millennium.

As to overpopulation, for example, they can try to devise harmless and acceptable methods of birth control—not at all an easy task, considering how much of an organism's physiology and behavioral repertoire is devoted to the propagation of its kind. But supposing them to be successful, they will have no special skills for solving the subsequent political, administrative, and educational problems of bringing these contraceptive measures into use among peoples who cannot read hortatory pamphlets, are not well used to taking precautions, and may anyway want to have as many children as possible.

Again, what can a scientist as such do about interracial tensions? Here his function is more likely to be critical than political; he will expose, maybe, the preposterous pretensions of racism and the whole farrago of genetic elitism that grew out of the writings of wicked old Sir Francis Galton. He may in the end convince political wrongdoers in the domain of race relations that they must not look to science to uphold or condone their malefactions. There are, in short, innumerable ways in which scientists can work for the melioration of human affairs.

The functions of a social mechanic or critic might be thought by many scientists to diminish their own—and science's—standing in the world. These would be mean-minded sentiments, though, and scientists will lose the influence they ought to and can exert if their pretensions are too grand or the claims they make for the efficacy of science exceed its capabilities.

The role I envisage for the scientist is that which may be described as "scientific meliorism." A meliorist is simply one who believes that the world can be made a better place ("Ah, but what do you mean by better?" and so on, and so on) by human action wisely undertaken; meliorists, moreover, believe that they can undertake it. Legislators and administrators are characteristically meliorists, and the thought that they are so is an important element of their personal *raison d'être.* They realize that improvements are most likely to be brought about by identifying what is amiss and then trying to put it right—by procedures that fall short of transforming the whole of society or recasting the entire legal system. Meliorists are comparatively humble people who try to do good and are made happy by evidence that it has been done. This is ambition enough for a wise scientist, and it does not by any means diminish science; the declared purpose of the oldest and most famous scientific society in the world is no more grandiose than that of "improving natural knowledge."

The scientists I envisaged in the two examples given above were consciously engaged in practical, or "relevant," endeavors. But what about the many scientists who do what is wrongheadedly called "pure" research? Where do they get *their* satisfaction? Nowhere if not in the advancement of learning itself.

Jan Amos Comenius spoke for them all. He dedicated his *Via Lucis*[3] to the Royal Society of London for improving natural knowledge ("Blessings upon your heroic enterprises illustrious Sirs!"). The philosophy they were bringing to perfection would, he believed, procure "the constantly progressive increase of all that makes for good to mind, body and (as the saying is) estate." Comenius's own ambition, touching in its magnitude and

3. There were not many copies of *Via Lucis* (1668) in the world when E. T. Campagnac made the translation from which I quote (London: Liverpool University Press, 1938).

breathtaking in its audacity, was to work toward a *pansophia:* "to weave together a single and comprehensive scheme of human omni-science" of which the purpose was "nothing in fact less than the improvement of all human affairs in all persons and everywhere." Those with enough hopefulness in their makeup willingly go along with the belief of Comenius that the pursuit of universal learning "to be acquired and applied to the benefit of all men for the common good" is truly *via lucis,* the way of light.

Index

About the Author

Sir Peter Medawar began research in Professor H. W. Florey's laboratory in Oxford in the early days of the development of penicillin. He subsequently investigated the causes of and sought a possible remedy for the rejection by human beings of organs and tissues transplanted from other human beings. For this research he was awarded the Nobel Prize in 1960, and his personal research work has extended over the entire field of experimental pathology. He was Director of the British National Institute for Medical Research from 1962–1971. He is currently in research on tumor biology in the Clinical Research Center of the Medical Research Council. His previous works include *The Art of the Soluble* (1967), *The Hope of Progress* (1974) and, in collaboration with his wife, J. S. Medawar, *The Life Science* (1977).